SUSTAINABLE ENERGY MANAGEMENT

SUSTAINABLE ENERGY MANAGEMENT

Planning, Implementation, Control, and Security

Second Edition

MIRJANA RADOVANOVIĆ
Professor of Energy and Environment, Educons University,
Sremska Kamenica, Republic of Serbia

ELSEVIER

ACADEMIC PRESS
An imprint of Elsevier

Academic Press is an imprint of Elsevier
125 London Wall, London EC2Y 5AS, United Kingdom
525 B Street, Suite 1650, San Diego, CA 92101, United States
50 Hampshire Street, 5th Floor, Cambridge, MA 02139, United States
The Boulevard, Langford Lane, Kidlington, Oxford OX5 1GB, United Kingdom

ISBN 978-0-12-821086-4

For information on all Academic Press publications
visit our website at https://www.elsevier.com/books-and-journals

Publisher: Charlotte Cockle
Acquisitions Editor: Peter Adamson
Editorial Project Manager: Sara Valentino
Production Project Manager: Sojan P. Pazhayattil
Cover Designer: Vicky Pearson Essar

Typeset by STRAIVE, India

Contents

Acknowledgments

This book is part of the Jean Monnet Erasmus + project titled "The circular economy: 'the number one priority' for the European Green Deal," Reference Number: 619927-EPP-1-2020-1-BG-EPPJMO-PROJECT.

The book was translated by Mrs. Sanela Šipragic Đokić.

CHAPTER 1

Introduction

Sustainable development is a comprehensive concept of development that was adopted in order to preserve the planet's resources to the extent that will satisfy the needs of present generations without compromising the ability of the next generation to meet the same needs. The concept is comprehensive, multidimensional, and multidisciplinary, and it can be applied widely.

The study of sustainable development involves considering the whole range of relevant issues, mechanisms, solutions, and critical attitudes. Currently, there is no single concept of sustainable development that can be universally applied in all areas of human activity throughout the world, and it is necessary to examine it in some sectors that are of special interest for a certain range of issues.

Since the creation of contemporary mankind, especially with the progressive development of human activities, there has been a need for the provision and consumption of certain types and amounts of energy. Every activity is related to the pending charges of energy that are created, transferred, transformed, and consumed.

Energy production and consumption are a special problem in the modern world, and it needs to be understood on several grounds. First of all, with a proportional increase in population, there is an increasing need for the provision of sufficient energy. Given the nature of population growth and inability to stop this growing trend, the problem of providing energy for all human needs becomes global and constant (Radovanović, Dodić, & Popov, 2013).

Energy needs are very different in different parts of the world. The highest energy consumption, since the beginning of the Industrial Revolution in the late 18th century, has been found in the most developed countries. Consumption of large amounts of energy has brought these countries a significant economic advantage and created the need to face a constant requirement of providing sufficient energy for future economic and social development (Arto, Capellán-Pérez, Lago, Bueno, & Bermejo, 2016).

Energy consumption in the world can be defined and measured on several grounds, but the most suitable is the energy consumption in the total

Sustainable Energy Management
https://doi.org/10.1016/B978-0-12-821086-4.00001-5

amount of time, in some regions, as well as energy consumption from particular sources.

The total energy consumption in the world undergoes complex monitoring and control, which has multiple objectives. First of all, monitoring of energy-related indicators in the world is necessary because it provides insight into the general situation of the energy sector, helps identify the differences among particular regions, as well as understand the trends and predict future tendencies. Tracking energy-related indicators requires extremely complex measurements, so that accurate data can be obtained only after several years. The framework overview of energy consumption in the world until 2010 with projections up to 2050 is shown in Fig. 1.1.

Energy consumption has been measured in an organized and precise way since 1965. Power consumption in the world is characterized by three periods. In the base year 1965, energy consumption in the world was less than 5 TW. Out of the 5 TW, the largest portion was provided by exploitation of traditional sources (oil and coal), almost in equal ratios. Half that amounts of energy was obtained from gas, and the least amount was obtained by exploitation of water energy. Even then, 1% of energy was obtained from nuclear power plants (Paul, Michael, & Jim, 2015).

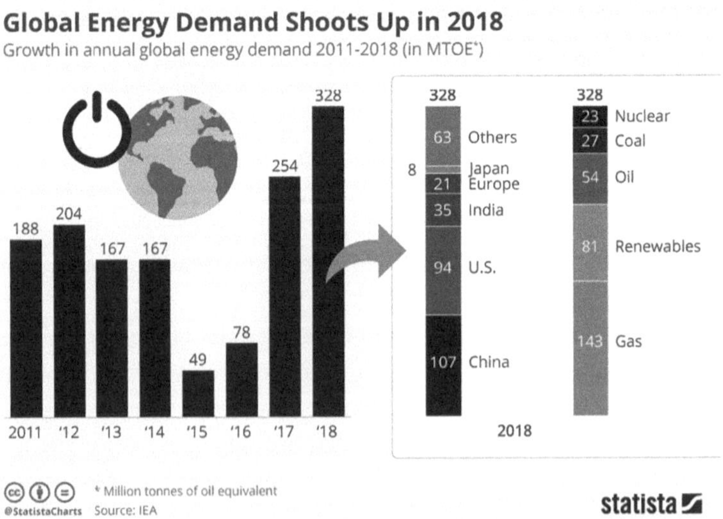

Fig. 1.1 Global energy demand. *(Based on Global Energy & CO_2 Status Report. (2019). https://iea.blob.core.windows.net/assets/23f9eb39-7493-4722-aced-61433cbffe10/Global_Energy_and_CO2_Status_Report_2018.pdf.)*

In the second period, which started in 1970, the energy consumption of petroleum was doubled and reached its maximum in 1980. Consumption of coal was also growing, but it was much slower and more balanced. Energy consumption increased steadily from gas, which can be partly explained by population growth and technology. The huge increase in consumption of petroleum can be understood only because of the increase in traffic. In this period (from 1980 to 1990), the consumption of water resources remained at the same levels, and the consumption of energy from nuclear sources grew slowly but not significantly.

The third period began in 1990 and is characterized by the beginning of a stable and balanced growth in energy consumption from traditional sources. Consumption of oil and gas continued to grow almost evenly, while the consumption of coal was stabilized and even slightly reduced by the end of the period. Production of energy from water and nuclear sources reached its peak in 1990, and there were no significant changes until 2000.

In the beginning of the new millennium, by a shift in the third distinct period, energy consumption from oil and gas continued to grow at the same pace, while the consumption of energy obtained from coal, after a decade of stabilization, began to grow again. At the end of the third period, in 2019, for which the precise data is available, the total world energy consumption was less than 15 TW, three times more than what was recorded 40 years earlier (World Energy Outlook Report, 2019).

The reasons for increased consumption and related predictions are numerous and complex and include the increase in the number of people in the world, improvement of the needs of industry and transport, inefficient consumption, and poor energy efficiency indicators. Whatever the reasons, this large consumption of energy, apart from the exhaustion of the world energy resources, has led to significant environmental consequences. All of which raises the need for a strategic redirecting in the field of energy management at all levels, with the knowledge that energy consumption is uneven in some parts of the world, as shown in Fig. 1.2.

Along with the monitoring of the total consumption of energy in the world, the monitoring of energy consumption in some regions of the world is connected with a number of problems, so final estimates cannot be considered absolutely accurate, although they provide valuable information about the approximate values of energy consumption.

The highest energy consumption can be seen in the United States, Canada, Norway, and Saudi Arabia, which can be largely explained by the high level of technological development and living standards, as well as the

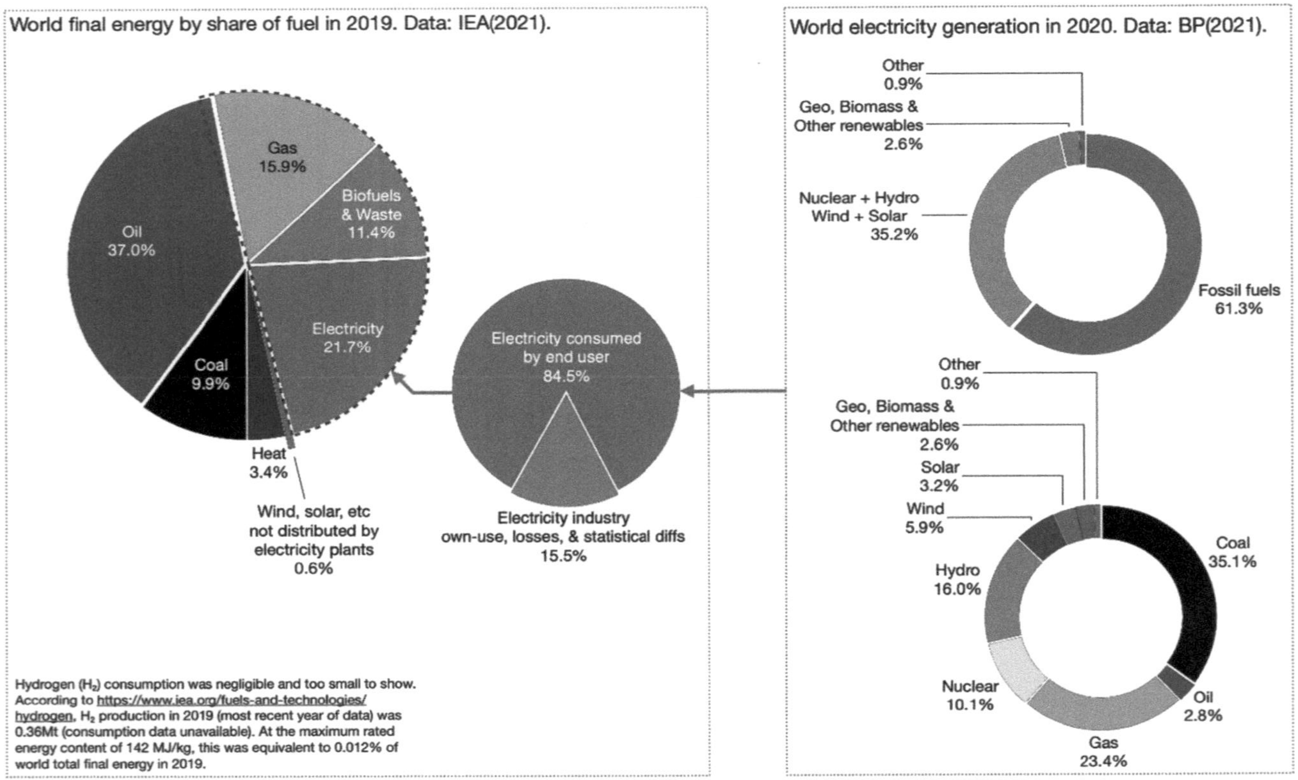

Fig. 1.2 World Energy Consumption in 2021. *(Based on World Energy Data. https://www.worldenergydata.org/world-final-energy/.)*

intensity of traffic. Somewhat weaker consumption has been recorded in Russia, Scandinavia, and Australia. Compared with the first group of countries, with the highest energy consumption (indicated above), the European countries, Japan, South Africa, and Argentina spend half that energy, followed by the Eastern European countries. Even lower power consumption has been recorded in Brazil, and the lowest in Africa and the Far East.

Energy consumption in some regions depends on many factors, but it is evident that it is related to the question of whether the region also produces the energy. Countries with the largest energy reserves are also the biggest consumers of energy, which cannot necessarily be considered justified. The European countries are mainly well-developed countries, with higher values of energy efficiency, which is mainly due to the great efforts for rational management of energy (Fawcet, Rosenow, & Bertoldi, 2019).

Consumption of energy per sector also has its particularities.

From the beginning of the industrial revolution (technical and technological improvements and expansion of transport), energy from oil is mostly used, and one-third of the world's energy needs is provided in that way. Apart from this, the trend of oil consumption is obvious. Over the years, oil consumption has continued to grow and that trend is expected to continue. Somewhat less energy is obtained by exploitation of coal (25%), but that trend has decreased, which can be explained by specific technological processes and the use of automobiles that cannot use coal as fuel. About one-fifth of world energy demand is obtained by the exploitation of gas (World Energy Outlook Report, 2019).

The remaining quantities of energy are provided in the same proportion of energy from the biomass (11%), nuclear (6.4%) energy, and water courses (21%). Exploitation of nuclear energy is limited to countries where such production is allowed. The least quantities of energy are obtained from alternative sources (2.2%). The average general values have been presented, but it should be emphasized that there are significant differences in certain countries and regions (World Energy Balances, 2020).

Taking into account the energy dependence of developed countries and the uneven distribution of energy resources, production and distribution of energy have become a problem of particular international political and economic importance. Energy in all its forms has become a subject of international trade, numerous disputes, negotiations, and military conflicts. Since the demand for energy on the planet is constantly growing, it will occupy a more important place in global economic and political changes in the future.

With the development of awareness and responsibility of individuals and businesses and countries with the objective to respond to the growing number of environmental problems, the concept of sustainable development has been defined to seek options for successful solutions to the environmental problems, including energy exploitation and consumption, which are connected with substantial negative environmental impact. The problem of providing sufficient energy is one of the main challenges of sustainable development and clearly breaks the trend of uncontrolled energy consumption and implicitly imposes the need for changes in this field.

Sustainable development involves, among other things, the gradual implementation of specific measures to drastically change the current approach to energy production and consumption, because it implies developing new technologies, utilization of new energy sources, development and implementation of comprehensive measures of saving, and the development and implementation of numerous legal regulations, all with the aim of raising levels of energy efficiency, i.e., to stop the trend of uncontrolled consumption of energy that inevitably leads to the rapid and total depletion of existing energy resources (Lindenmayer, 2017). These measures are binding for all countries that accept the concept of sustainable development and are a part of many international agreements and protocols that govern this area.

The energy problem, with the need for economic development on the one hand and the need to keep energy resources and reduce pollution on the other, requires definition of a special mechanism to manage this area, which is essential to the process of planning at all levels, in companies, across regions and countries, and in the international community as a whole (Romero, Linares, & Lopez, 2018). Sustainable energy management, with all the elements of modern management science, along with the integration of requirements that are set by implementing the concept of sustainable development, is an effective mechanism that allows long-term planning of sustainable development. Sustainable energy management is designed so that its steps are clearly defined and measurable and its goals are specified and measurable. Only this model of management, which is based on the impact of the factors of sustainability, with clearly defined goals, provides quality monitoring of the progress of the whole process of sustainable development. However, in sustainable energy management, the impact that energy exploitation and utilization of energy resources may have on other indicators of development can be monitored, measured, and kept under control.

Sustainable development consists of four subsystems that are interconnected although different in nature and intensity. The economic, environmental, social, and institutional subsystems are a kind of sustainable development unit, whose harmonious development represents the best possible option for long-term, stable development and survival of mankind. Energy management is a key problem of sustainable development because energy consumption reduces the values of the environmental subsystem but has positive effects on the values of the economic subsystem.

Sustainable energy management enables the monitoring of economic and ecological parameters of development because it includes the exact methods for assessing the effectiveness of implementation, and the control is based on a method for determining the degree of sustainable development through measuring its indicators. Defined in this way, the sustainable energy management model can provide real insight into the level of achieved sustainable development as a whole, taking into account that determining the balance between the economic and ecological development is particularly important (Siraganyan, Perera, Scartezzini, & Mauree, 2019).

Sustainable energy management is designed to provide a management model that can be accepted not only by individual companies but also by countries or regions that create a geographical, natural, and economic unit. The sustainable energy management model is based solely on determining the most appropriate form of management of the system of future energy production and consumption, indicating the most important objectives in the field of sustainable energy development, pointing out how these goals can be achieved, and controlling the complex process by applying the methodology of assessment of the achieved degree of sustainable development. Only in this way is it possible to plan, organize, monitor, and control the complex and complicated process of energy management in the modern world.

References

Arto, I., Capellán-Pérez, I., Lago, R., Bueno, G., & Bermejo, R. (2016). The energy requirements of a developed world. *Energy for Sustainable Development, 33*, 1–13.

Fawcet, T., Rosenow, J., & Bertoldi, P. (2019). Energy efficiency obligation schemes: Their future in the EU. *Energy Efficiency, 12*, 57–71.

Lindenmayer, D. (2017). Halting natural resource depletion: Engaging with economic and political power. *The Economics and Labor Relations Review, 28*(1), 41–56.

Paul, E., Michael, B., & Jim, W. (2015). *Global energy: Issues, potentials, and policy implications.* Oxford University Press.

Radovanović, M., Dodić, S., & Popov, S. (2013). *Sustainable energy management* (1st ed.). Elsevier.

Romero, J. C., Linares, P., & Lopez, X. (2018). The policy implications of energy poverty indicators. *Energy Policy, 115*, 98–108.

Siraganyan, K., Perera, A. T. D., Scartezzini, J. L., & Mauree, D. (2019). Eco-Si A parametric tool to evaluate the environmental and economic feasibility of decentralized energy systems. *Energies*, *12*(5), 776.
World Energy Balances. (2020). *International energy agency*.
World Energy Outlook Report. (2019). *International energy agency*.

CHAPTER 2

Energy and sustainable development

Contents

Sustainable development as a theory, concept, and idea is a comprehensive framework for the development of humankind in the future, i.e., the attempt to plan future development based on past experience and predicted future needs. The concept of sustainable development implies the need for reviewing and understanding its complex and multidisciplinary, multidimensional, and heterogeneous structure, making it one of the most complex concepts of development since the creation of humankind.

Any human activity can be observed from different points of view and evaluated on several grounds. Sustainable development, however, being the starting assumption and the basic criterion for the acceptance of the idea, requires that the observed activity, event, or asset at any stage of its implementation, development, and consumption not have any detrimental effect on the environment. Broadly, sustainable development is the need to encourage and allow the performance of only those activities that do not have a detrimental impact on the quality of life of future generations.

Sustainable Energy Management
https://doi.org/10.1016/B978-0-12-821086-4.00002-7

Taking into account the different types of human activities, diversity of resources, and ways of their exploitation and consumption, it is necessary to study certain aspects of sustainable development (economic, environmental, and social) as a separate phenomenon, but also consider its relationship with all other parameters of sustainability.

Definition of sustainable development

Sustainable development is a recently accepted concept of development that has emerged as a result of the fact that the development of civilization has exhausted natural resources to the extent that the earth has become unsustainable, thus challenging the prospects of development and survival of future generations. The awareness of the need for preservation and restoration of natural resources has always existed, as indicated by written documents and oral traditions. In the early periods of human development, people were much more aware that they depended on the earth and they treated it with respect. With the increase in population, geographical discoveries, the exploitation of colonies, and the Industrial Revolution, people suddenly stopped being interested in the preservation of the human connection with the earth, and its resources have been exhausted relentlessly in the last two centuries. Only after World War II, especially after the Chernobyl nuclear accident (Ukraine, 1986), was public awareness raised again about existing and future environmental problems (Environmental consequences of the Chernobyl accident and their remediation: Twenty years of experience, International Atomic Energy Agency, Vienna, 2006).

The first public reaction prompted authorities to start to legally regulate pollution, and in the 1980s, environmental issues were raised to a higher level, and the emphasis shifted from sanctioning polluters to prevention of the problem.

Laws and technology have prescribed and found a variety of solutions that prevent environmental incidents or decrease their likelihood. The scope of interest has also extended to the field of ecology, and the need for conservation, rational exploitation of natural resources, and replacement of non-renewable resources have become imperatives of environmental protection.

Sporadic solutions and intentions of certain countries to make progress in the field of ecology soon encountered the barriers that exist in the form of borders, social organizations, and conflicting economic interests, but global environmental problems have been increasing, and the international community is aware that it has to cooperate in this field. The need to define a new concept of growth and development of the planet is the imperative that would determine further development of humankind (Andrews, 2020).

The concept of sustainable development was first officially used as a possible model of development in 1987 at the 42nd United Nations General Assembly in the report of the Commission for Environment and Development called "Our Common Future." The Commission was established in the 38th session of the General Assembly in 1983, and the report was widely known as "The Report of the Brundtland Commission," named after Gro Harlem Brundtland, the Norwegian Prime Minister who chaired the historic session. The most commonly cited definition of sustainable development is the definition given in the Brundtland report: "Sustainable Development should meet the needs of present generations without compromising the ability of future generations to meet their needs." If development is defined as an increase in welfare, then sustainable development can be understood as preservation of welfare over time, the Brundtland commission report: Our common future (Oxford University Press) (1987).

The Brundtland Commission promoted the politically correct idea of sustainable development, and its report is similar to the Club of Rome report of 1972, which was published under the title "The Limits to Growth." (Meadows, 1974). In contrast to Brundtland's report, "The Limits to Growth" was sharply criticized, especially by economists, and did not have a significant influence on international and national environmental policy. This report was seen as highlighting the importance of redistribution, while Brundtland's report was interpreted as a desirable continuation of economic growth. Brundtland's report became notable due to four important observations, as follows (Report of the world commission on environment and development: Our common future, presented to the UN general assembly in 1987, 1987):

1. First, it proposed the concept of sustainable development, defining it as meeting current needs without compromising the ability of future generations to meet their needs as well.
2. Second, it emphasized that international cooperation is essential but very difficult to achieve. Brundtland's report points out the problem succinctly: "The earth is one, but the world is not." The policies of different countries are often driven by local and regional interests that do not comply with global environmental expectations.
3. Third, it recommended adoption of the United Nations program on sustainable development and organization of the International Conference on Environment and Development.

4. Fourth, it suggested the strengthening of national environmental agencies, institutions, and organizations.

Thus, the concept of sustainable development was widely adopted quickly. During the United Nations Conference on Environment and Development, popularly called the Earth Summit, held in June 1992 in Rio de Janeiro, Agenda 21 was defined and ****adopted as a comprehensive plan of action for achieving sustainable development in the 21st century. In 2003, the World Summit on Sustainable Development was organized (Rio 10+), and the importance of sustainable development was reaffirmed by individual countries and globally. It was also acknowledged that the implementation of the idea of sustainable development had not advanced significantly, and therefore, it was necessary to take concrete actions and measures to fight poverty, change unsustainable models of production and consumption, and protect rational provision of natural resources (Agenda 21, United Nations conference on environment, & development Rio de Janeiro, Brazil, 3 to 14 June, 1992).

One of the first definitions of sustainable development was given by Repetto, who said that the idea of sustainability is basically grounded in the belief that decisions made today should not jeopardize the prospects for the preservation or improvement of living standards in the future (Repetto, 1988).

This definition implies that economic systems should be managed by using dividends on available resources, where the resource base should be kept and improved, so that next generations as well as the previous ones can live. This view has many similarities with the ideal concept of income considered by Hicks more than half a century ago. According to Hicks, the calculation of income makes it possible to determine the maximum amount that can be spent in the current period, while at the same time not diminishing the prospects for consumption in the future (Hicks & Simon, 1979).

More recently, Solow defined the concept of sustainability—concept of elasticity. According to Solow, sustainability does not imply an obligation to spend natural resources according to sustainable needs. It is necessary to substitute all nonrenewable natural resources with renewable alternatives. Also, the pattern of production consumption should be reconsidered and redefined in order to spend optimal quantity of resources and to produce less possible waste (Solow, 1993).

The principle of sustainability highlights the existence of freedom to use resources that future generations will be deprived of as long as their living conditions remain the same as the conditions provided for present generations. In other words, inability of future generations to use resources would mean a departure from the criteria of sustainability, if that worsens the living conditions of future generations as compared with the previous ones. Although sustainable development may be understood and interpreted differently, the problem occurs when it is necessary to formulate a model of sustainable economy.

Ten years after the Brundtland's report, determination of sustainable development became globally recognized and the subject of discussion on the highest level started together with its global challenge (Conference in Rio de Janeiro in 1992). Therefore, sustainable development started to be connected with the need for certain changes at all levels (local, regional, and global), plus intensification of international cooperation, strongest commitment, and awareness of responsibility for well-being of future generations.

Although this concept was accepted in 1987, there is no proposal for its operational implementation.

Only in recent debates on sustainable development has the idea of a long-term preservation of nature been discussed. In the past, nature was seen as something needed to meet current and future needs, and the idea of sustainability was not environmental, but primarily economic.

As a whole, the concept of sustainable development emphasizes the international aspect of economic activities and their impact on the environment and future resources. Therefore, the opportunity (direct and environmental cost) of providing a certain future is in the center of the concept. Sustainable development is connected with the concept of optimal growth. It aims to establish a balance between economic growth and environmental degradation. The costs of environmental pollution and utilization of resources have an important role in making current economic decisions, since they represent the cost of the damage inflicted on the environment and natural resources that would be paid by future generations.

The concept of sustainable development includes, apart from economic issues, social aspects such as poverty, social disorders, and other problems of social and political stability. It also includes degradation, environmental pollution, and depletion of natural resources. Therefore, sustainable economic development could mean an increase in real GDP (Gross Domestic Product), basic indicator of economic development per capita, which is not the result of environmental degradation, exhaustion of resources, and social disorders. In this interpretation, the environment and nature are designated as a certain restriction on economic development. The concept of sustainable development seeks to prevent negative economic growth that may increase future costs (Matthews, Allouche, & Sterling, 2015).

The normative content of economically sustainable development is defined in the requirements to preserve the natural conditions and environment for future generations. The implicit requirements of the concept of sustainability are the following (Sauvé, Bernard, & Sloan, 2016):

- Sustainable intergenerational and economic wealth of people;
- Ensuring the survival of humankind as long as possible;
- Striving for flexibility in production and economic systems and/or stability of their characteristics (their ability to recover when exposed to shocks);
- Maintenance of biodiversity;
- Ensuring sustainability of the community; and
- Stabilization of the biosphere.

The complexity of environmental issues, which are primarily a consequence of the rapid development of Western industrialized countries, has influenced the acceptance of the notion that traditional economies cannot resolve these questions adequately. If the volume of economic output increases (being the result of the increased consumption of energy and raw materials, as well as the growing number of people), this would necessarily lead to greater use of natural resources and environmental services.

Sustainable development principles

While some authors believe that the concept of sustainable development is a new theory or doctrine of development, most believe this is a new concept whose wider implementation would change patterns of production, consumption, and everyday life, because it is partly based on the new generation of normative ethical theories. This concept is a landmark ("asymptotical ideal"), and the paradigm of sustainable development is never directly applicable. In order to be realized, the sustainable development concept requires undertaking of new activities as follows (Roche & Campagne, 2017):

- New legal and institutional arrangements;
- Adequate financial and economic arrangements;
- New technology and technical solutions;
- Promotion and education;
- New methods of public communication and interaction, with an emphasis on openness, participation, and transparency; and
- New coalitions for sustainable development.

However, these social, economic, political, ecological, spatial, and intergenerational principles and criteria are still general and therefore are not directly applicable. They require operationalization, which has been undertaken in most countries facing this problem. The specific strategy or policy of sustainable development combines a number of general and specific principles and criteria. Depending on the number, content, and specific problem they relate to, the following general principles and criteria of sustainability are usually partly operationalized (Scoones et al., 2020):

- The principle of strict conservation and the principle of precaution. This principle gives priority to environmental protection. This is especially applicable in situations where scientific and other objective views do not provide sufficient knowledge on potential consequences of the anticipated economic activities, so the rigorous conservation of natural resources should be exercised even at the cost of limiting economic growth. However, applications of these principles should not prevent the use of innovative procedures that can be easily predicted to have negative impacts on the environment.
- In the analysis of environmental impact and costs and benefits, the central place should be given to the principle of alternative solutions, it being one of general principles. When more intensive exploitation of natural resources is inevitable, it should: (1) be minimized, (2) use nonrenewable resources as little as possible and renewable resources as much as possible, and (3) be applied with respect to high and rigorous ecological and spatial standards. Intensive economic growth that would be accomplished at the cost of significant exploitation and endangering of natural resources, especially the nonrenewable ones, is not acceptable.
- The principle of rational use of raw materials and other resources and inputs.
- The principle of risk reduction is applied when there is an estimate and economic activity can be understood as an objective danger and risk to the environment.
- Applying the principle of durability (of products and services) allows the production of quality output, with reasonable costs, which are socially and market acceptable.
- The principle of "polluter pays" and the principle of "user pays" are applied in all cases where economic activities threaten the ecological capacity and end thresholds.
- Depending on the type of activities and types of areas, the following principles are applied individually or in combination: (1) planning

principles; (2) market principles; and (3) principles of regulatory control (bans, permissions, etc.).

Application of the principle of reusability/using and recycling should lead to expanding the use of resources that can be reused or that are biologically degradable. Decisions on projects, programs, and plans for economic development and protection should be made based on the analysis of their impact on the environment. In addition to the analysis of direct costs and benefits, the main impact of indirect costs and benefits should be evaluated. Market prices as parameters in evaluating the impact should be used if the assumption on the smooth operation of ("perfect") market mechanism is justifiable. In cases where this assumption cannot be justified, the parameters from the class of "shadow price" should be used, etc. (Jha-Thakur & Khosravi, 2021).

Economic decisions should be based as far as possible on sound systematic insights into the ecological-spatial capacity and extreme thresholds, as a general reference and basis for evaluation.

The principles of public openness, democracy, and participation should be the basis for the preparation, adoption, and implementation of decisions on sustainable economic development and environmental protection, with appropriate IT support.

Practice has shown that general principles and categories of sustainability are not directly usable in preparation, adoption, and implementation of business decisions and all decisions related to sustainability. These principles, therefore, need to be operationalized and generalized so that they express the criteria and the contents of a specific market and geographic space. The disputes over the various meanings of sustainability originate here.

While there is a kind of consensus about general principles and criteria in terms of selected interpretations of sustainability, different interpretations can be found in cases when concrete goals have to be determined, as well as the means for their implementation. In such cases, certain aspects of sustainability are emphasized and some other aspects are ignored. In such division, the aspects of interest play the most important role. On the other hand, this is a consequence of the fact that sustainability is a multilayered phenomenon that has many meanings that can be mutually incompatible. In practice, there are very few situations where it is possible to implement all layers of sustainability, which indicates the necessity to make a "trade-off." Given that sustainability has several meanings and contexts, for community understanding, the following defines sustainable:

- Something that provides continuous economic growth and development;

- Something that is socially acceptable;
- Something that harmonizes the ratio and the measure of economic and social development according to ecological capacities; and
- Something that is politically acceptable (can be the subject of agreement in official institutions).

Among the various meanings and contexts of sustainability, economic, social, and political sustainability emphasize the importance of so-called intragenerational equality and fairness, while intergenerational equality and fairness are of greater importance in the ecological understanding of sustainability.

Energy sustainability as criteria for development

A particular problem today and a challenge in the field of its development is how to define the terms sustainability and acceptability in terms of sustainable development. Sustainability can be understood as a term, a concept, a philosophy, an attitude, a request, or a need. The current concept of sustainability has emerged as a result of decades of defining activities and events that are acceptable or unacceptable in terms of opportunities to meet the needs of present and future generations. A particular problem is the need to define the concept of sustainability when it comes to energy development and security in this area (Lior, Radovanović, & Filipović, 2018).

In the early 1970s, the first significant energy problems struck the world's social scene, and the first ecological requirements were generated. Later, ecological regulations and recommendations were standardized and continue to be developed today. With the development of ecology as a multidisciplinary science and an increase in people's awareness of the need to preserve and improve the quality of the environment, the term "sustainable" has begun to be widely used.

At first, ecology, and thus sustainability, was seen narrowly as a set of rules that exist just to prevent environmental pollution. In the 1970s, the first regulations in the field of ecology appeared, but all remained limited to the regulation of physical, chemical, mechanical, and microbiological impacts of some agents on the immediate environment.

During this time, with the rise of awareness regarding energy problems and the increasing number of companies looking to gain a profit at the environment's expense, sustainability took on a new dimension. It was at this time that government agencies started making reports about the deterioration of the environment and the depletion of nonrenewable energy resources. The public became familiar with the problems in the environment and demands emerged for conflict resolution. It was then that ecology was given a new dimension, and many intensive studies about the degree of exploitation of renewable and nonrenewable natural resources began.

This awareness resulted in the adoption of the concept of sustainable development at the highest level of decision-making as the only prerequisite for long-term quality survival. Because of this, sustainability has been given a new dimension.

In the early 1980s, it was suggested that a sustainable energy process is one based on the accepted concept of sustainable global development and the development of such business and any other strategy that adopts the criteria of the concept. During this early period, the impact of political structure and citizens was intensified, and relevant laws, regulations, recommendations, proposals, and numerous other initiatives were adopted. The period of the early 1980s can be seen as a turning point in understanding and respecting the criteria of sustainability at all levels. Developed countries were leading in this movement. Previously adopted regulations on the prevention of pollution of nature were tightened and new ones were created. Due to the development of other sciences, such as chemistry, biology, and medicine, new findings were made on the harmful effect of certain substances in nature, and the list of hazardous and potentially hazardous materials and activities was expanded. Sustainable became only that which was not on the list of activities that threaten the environment. Later, the rejection of certain energy processes started taking place, which later led to the closing of whole power plants that were not environmentally suitable (Gunnarsdottir, Davidsdottir, Worrell, & Sigrugeirsdottir, 2020).

Sustainability of energy at that time moved out of the domain of simple monitoring and sanctioning the work of power plants. New strategies were adopted to preserve the quality of environment. Long-term sustainability became an important factor in decision-making processes when planning new power plants or infrastructure used for the transfer of energy. Ecological movements were strengthened along with political parties that observe ecology as a priority. In such an environment, energy sustainability was given new importance, as the living environment began to be seen as unique (global), and people began to accept themselves as inseparable parts of nature.

Human activities that did not adversely affect nature and man himself became sustainable, which led to the expansion of savings of all types of materials, the study of alternative forms of energy, and the problem of nuclear power plants was raised. Companies that were not ecologically suitable came under attack, paid huge fines, and were boycotted or closed forever.

Nothing can be observed separately from the living environment as a whole. Science is increasingly trying to study and explain the complexity of interactive processes in nature, and man understands that any activity that he undertakes will, after a certain period of time, affect his health and quality of life in general.

It is quite clear that the energy sustainability of a particular process or activity can no longer be regarded only as a simple set of numbers and standards. Most people cannot (and do not have to) explain the harmful impact of a process but can clearly see the consequences. People want not only to avoid such impacts but to prevent them. Awareness about limited energy sources and need for their replacement for the first time becomes part of strategies for development of companies, development of the region, and the global community as a whole.

In the 1990s, energy sustainability ceased to be viewed only as a precondition for the preservation of the environment. Instead, modern man accepted the fact that economic suitability should be observed in the broadest sense possible. Sustainability ceased to be seen as a qualitative characteristic of a process, but became a parameter of evaluation in all areas of life and work. Environmental requirements in the broadest sense became something that must be respected in order to survive and make progress.

Also, in the 1990s, sustainability began to be widely applied in the description and evaluation of not only economic activities but also in the field of politics, culture, education, sports, and entertainment. The environmental impact of any human activity was examined, and at the end of the last millennium, developed countries raised awareness of the need for ecological evaluation and exclusion of anything that can negatively affect humankind. The principle of responsibility toward the environment, toward us, and others became an integral part of cultural values, especially in the most developed countries (Luna-Nemecio, Tobón, & Juárez-Hernández, 2020).

The new millennium brought about new challenges and problems, but the desire of the international community to make ecological improvements whenever possible is clear. Sustainability, and therefore the sustainability of energy, has become a philosophy of life, and each individual is educated to support the requirements to live a healthy and quality life and to be aware that generations to come have the right to have such a life as well. Sustainability is not easy to achieve—it has to be a part of laws and a generally accepted human value. Energy sustainability is no longer desired, it is a must. Of course, most developed countries have already achieved this level of sustainability, but the situation is much different in underdeveloped countries (Filipović, Radovanović, & Golušin, 2018).

Dimensions of energy sustainability

In order to more easily accept energy sustainability as a business philosophy, it is first necessary to determine what is desirable in terms of environmental protection and what should be strived for so as to direct the activities in that way. Defining the concept and the essence of energy sustainability is a particular challenge because the list of ecologically suitable or unsuitable activities or phenomena can never be final.

Energy sustainability, especially in developed countries, has been broadly analyzed, so that it can freely be said that energy sustainability can be sought in any phenomenon and activity that takes place in the sphere of production, transmission, and consumption of energy. Energy sustainability, contrary to many qualitative measures, has a kind of temporal, spatial, political, economic, and civilization dimension (Lior & Radovanović, 2017). It is hard to say that the dimensions of energy sustainability are determined once and for all and not subject to changes. Time will inevitably bring many new requirements, but it is still necessary to accurately determine the specific features of these dimensions.

The *time dimension* of energy sustainability implies that the time factor is very important in ecology and includes several methods of observation of the phenomena in the field of energy, which can be described as sustainable or unsustainable. First, an observation is made, as well as the study and proper assessment of events from the distant and recent past, in order to determine their impacts on the current situation and learn useful lessons.

Monitoring and controlling all current phenomena rely directly on the past and most directly on the future. Naturally, energy sustainability means conducting appropriate activities now and understanding how they will affect the state of energy in the future because it includes what is most important in sustainable development—the timely prevention of all ecologically unacceptable events and activities. Future may be limited only partially and for practical reasons, because the whole concept of sustainable development is based in the long term on the principle of intergenerational justice.

The *spatial dimension* is based on the generally accepted fact that energy sustainability can and must be observed in any space. Each individual can and should find sustainable and environmental-friendly behaviors and actions in their immediate vicinity and according to that initiate their wishes and requirements. Ecological eligibility is observed, found, and presented at the level of settlements, areas, countries, regions, and so on. Of special importance is the understanding that energy sustainability should be viewed primarily from the global point of view.

The *political dimension* has become particularly important in the past two decades. Numerous changes in the political arena and misunderstandings and conflicts are associated with the growing insecurity in the field of energy security of countries and regions. All developed countries in the world estimate sustainability of energy, both in their own country and the region. Because of that, intensification of political interference can be expected in determining energy sustainability and all it implies in the near or distant future. Respecting the rules of sustainable development has become one of the prerequisites of integration in world trends.

The *economic dimension* of energy sustainability is of particular importance, given that successful economic development and stability are of immediate relevance for each individual and are closely linked with the capabilities to meet the needs for energy. Today, businesses face great challenges as well as opportunities to respond to consumer demands that send clear messages that they do not want to support any unsustainable energy or environmentally harmful technologies and products. In most developed countries, there is no special requirement for educating citizens in that direction, given that the area is, according to public pressure, fully embedded in the institutions of the system.

The laws are strict, efficient, and apply equally to all. Business processes are reviewed in developed countries even before they have started to determine their ecological suitability. Substantial financial resources are allocated for that purpose. Companies that plan to operate in the future have to accept such rules, not only because of legal regulations, but even more because of the demands of consumers.

The *civilization dimension* of sustainability is a kind of summary of all the above and includes much more, so it is particularly difficult to define and almost impossible to restrict. Thus, energy sustainability must be considered together with the sustainable development of all other activities. Modern man has a right to live in a proper living environment with water, air, soil, flora, and fauna. However, most importantly, the human environment includes other people and their relationships. Physical and chemical parameters are relatively easy to determine, and it can easily be identified what is ecologically suitable and what is not and then make decisions as to what is ecologically suitable.

Inseparability of human nature and the inevitable development of human relations are the two main factors that determine sustainability in an entirely new way, which means that the actions of each person must be evaluated in terms of ecological suitability. Man is in constant interaction with nature, with other people, and with himself. Through thousands of years of history of humankind, experiences have been accumulated along with achievements that have become an integral part of every human being.

Human problems, misunderstandings, wars, and intolerance have strongly influenced the development of the consciousness of modern man. A civilized man should not base his existence on the disappearance and destruction of his environment to meet his own, often short and limited objectives. It is unrealistic to expect that man would give up a certain quality of life for ecology, but with a little more effort and cooperation, energy development in harmony with nature can be achieved, which will have only long-term positive effects on the mental and physical health of each individual.

Basic concept of energy sustainability

The international community as a whole is a special challenge. Differences in the degree of economic development, political system, culture, religion, and customs among different countries and regions have led to different ideas about energy sustainability. Therefore, in order to study these issues, it is necessary to briefly examine the historical development in certain parts of the world, given that the events in the past directly influence the current level of development of environmental awareness and therefore the understanding of energy sustainability.

Today, the most developed countries have the most elaborate system regarding managing the quality of the environment, as a direct response to strong public pressure. Thanks to a high degree of economic development, advanced education systems, and the achieved level of responsibility and political maturity of society, consumers openly express their demands, seek answers, and often opt for environmentally friendly technologies, products, or services. Legislation in the field of ecology is developed and functional, so it can be said that the systems in developed countries almost entirely support the concept of sustainable development. The European Union countries, Scandinavian countries, United States, Canada, and Japan are leading in terms of pointing out the importance of ecological suitability in all areas of life and work. Energy sustainability in these countries includes economical use of energy resources, energy-efficient operation, production of energy-efficient products, and use of renewable energy sources (Lacey-Barnacle, Robison, & Foulds, 2020).

On the other hand, there are a number of countries at the medium level of economic development, concentrated in southern and eastern Europe, which lag behind in terms of attention to ecology in general. The history of the development of these countries has not been favorable for the development of a healthy economic system, so ecology has been pushed into the background. Thanks to broad education, citizens are familiar with environmental problems, they support the concept of sustainable development, and from time to time highlight their demands, but generally poor conditions in the environment do not create a favorable ground for advancement in the field of ecology in underdeveloped countries. A society that is burdened with many problems is not able to fully address the problem of ecology and sustainable energy development, but companies need to be aware that their future survival, and above all possibilities of exports, depends on business operations that are in line with global environmental standards (Primc & Slabe-Erker, 2020).

The largest group of countries consists of nondeveloped and underdeveloped countries at a very low level of economic and social development. They are the victims of historical events and not strong enough to successfully handle their own existential problems, let alone pass appropriate laws and educate their citizens in the direction of ecological and energy sustainability. It is a paradox that these countries were examples of untouched nature and are rich in flora and fauna, as well as energy and mineral resources, where man lived on, in our terms, a low level of development. Today, these are the countries whose resources are maximally exploited with little or no environmental protection. The industrial plants of the so-called dirty technology and huge amounts of waste are moved to third-world countries. Inhumane working conditions, exploitation of child labor, poor health care, and the unavailability of broader education are just some of the indicators of the quality of life (Mirković-Isaeva, Radovanović, & Vasić, 2018).

Sustainable energy development is the need to harmonize the energy development of humankind in accordance with the possibilities—natural energy resources. Regardless of the comprehensiveness of the problems and diverse situations in the world today and that are expected in the future, several basic concepts of sustainable development have been defined, as follows (Ing, 2020):

- The concept of nonexhaustible energy resources includes conservation of both natural and gained total natural energy resources;
- The concept of nonexhaustible natural energy resources results from the previous concept and implies that natural energy resources are held

constant, considered as a whole and per capita, and should be especially considered because of constant population growth;
- The concept of elasticity, which explains the reduction of specific natural energy wealth, is necessary and inevitable and occurs as a logical consequence of population growth and growth of their demands.

It should be emphasized that these concepts were generated over time after the adoption of Agenda 21 at the UN conference in Rio de Janeiro in 1992. But certain assumptions about the basic concept of sustainable development existed even before, starting when the energy crisis of the 1970s initiated the change in the approach and the mode of exploitation of nature by man.

In the very beginning, there was an assumption, actually a requirement, to fully preserve energy resources as they were found. Consequently, this meant the cessation of further exploitation of energy resources and their complete conservation. In certain countries, and in some cases, the idea of total cessation of exploitation of energy resources was implemented, but it was sporadic and negligible. Given the rapid population growth and growing energy needs, the concept of nonexhausting and acquired natural wealth was abandoned.

Taking into account the shortcomings of the concept of nonexhausting and acquired natural resources, the concept of nonexhausting wealth was adopted and took into consideration the number of inhabitants on the planet. In fact, this concept implicitly implied that each inhabitant of the planet should be given a certain amount of energy wealth. This concept tried to eliminate the problem of the place of birth of the inhabitant. Actually, this concept implies that all-natural resources belong to all, regardless of their place of birth or residence. Regardless of the civilization progress, the concept was defined in such a way that could not be sustained in the long term, primarily because of the uneven distribution of energy resources on the planet and the lack of intention to introduce any changes in the field of energy production and consumption. Energy resources have already become an important element of world trade and the way of acquiring economic and social power.

Currently, the most acceptable concept is the one of elasticity, which takes into account the basic shortcomings of the two previously developed and abandoned concepts of sustainable energy development. The concept takes into consideration the global condition on the planet and above all the increase in the number of inhabitants and the need for a certain quality of life that the inhabitants are not willing to abandon. But there is a need to revise the current power management mode, both globally and locally.

Changes in the traditional way of thinking, doing business, and dealing with energy at all levels are needed.

In short, the concept of sustainable energy development, i.e., its basic principle from which all other principles are built, is based on the principle of intergenerational justice, which includes meeting the energy needs of present generations without compromising the ability to meet the power needs of future generations. In other words, the current generation has an obligation to leave to the next one the energy situation it had itself.

The basic concept of sustainable energy development can be developed based on different criteria, but it can be generally summarized as follows (Mallory & Ando, 2014):

- The principle of conservation of the existing nonrenewable resources, where, depending on the type, quantity, and quality of available resources as well as the needs and opportunities in specific areas, the most acceptable way of conservation of existing resources is considered, or permanent reduction of the intensity of their exploitation;
- The principle of exploitation of renewable energy resources, which to some extent replaces the use of nonrenewable resources and allows the use of energy in areas that are insufficiently rich in traditional energy resources, or in hardly accessible areas;
- The principle of energy efficiency, which implies an efficient and economical utilization of energy in all phases of its existence, from the energy that is accumulated in the resources through efficient production, distribution, and consumption, to promotion and support, to production of goods that use less energy than the same or similar goods;
- The principle of intergenerational justice, which incorporates the rule of energy management in all plans of energy development that would enable future generations to meet their own needs for energy;
- The principle of harmonization of economic development and energy consumption, which determines the development discussed above, is an especially sensitive principle whose implementation is linked with many problems, because it suggests the need to change traditional ways of thinking. In fact, economic growth was seen as the only measure of the progress of each country, and only indicators of economic growth were significant and available in determining the situation in each country and its position in the international community. The need for sustainable energy development imposes the acceptance of the new principle, by which it is necessary to cease economic growth if it leads to excessive depletion of energy and other resources;

- The principle of paying damages caused by excessive and inappropriate use of energy resources, which creates the possibility that all countries in the world and all business entities understand the importance of adequate energy management and will take on obligations resulting from that and responsibilities if they do not fulfill these obligations;
- The principle of measurability is derived from the need for sustainable strategic management, which is possible only if during the process of planning, implementation, and evaluation we deal exclusively with goals and data that are strictly and exactly measurable. Arbitrary and universal definitions, plans, and objectives do not provide a true picture and ensure proper application of sustainable energy management tools that are available; and
- The principle of promotion and education, which imposes the need for continuous promotion of sustainable development of energy. Only with thorough education that begins at an early age can conditions be created for the long-term responsible treatment of energy and creating a future that is not limited.

These basic principles of energy sustainability highlight the basic framework necessary for planning and implementation of sustainable energy management and present a framework within which energy management can be implemented. Only through the respect of the basic principles can energy stability and long-term energy sustainability be provided.

Basic problems of future energy development

The future of the planet carries a number of assumptions about future environmental problems, which the current generation faces and which future generations will face, too. Therefore, the main challenges to the concept of sustainable energy development are defined globally, as follows (Delina & Janetos, 2018):

- Demographic growth is one of the factors against which it is not possible to act globally with any measure. There are examples of countries that seek to encourage or restrict population growth by application of certain measures, but these are examples that cannot have a global impact. From the very beginning, the world's population has been steadily and progressively increasing. It would not take long to double the number of inhabitants on the planet. At the beginning of the 21st century, it took almost a thousand years to double the population, while current estimates indicate that the number of inhabitants on the planet will double in 50 years. Realistic estimates suggest that by 2025, about 8.4 billion people will be living on the planet. The significant part of the current population suffers from hunger, thirst, poverty, and unemployment. Demographic problems are characterized by the following:

- Global overpopulation, because obviously more people live on the planet than is optimal, and prospects in terms of that are not encouraging because the number of people in the future will increase;
- Uneven population density, as a characteristic of the modern world, which creates many problems. Globally observed, the population on the planet is very unevenly distributed as a result of development of transport and migration of population in search of better living conditions, as well as the consequences of uncontrolled high birth rates. There are areas in which extremely large numbers of people live in difficult economic and health conditions (certain regions of Africa, Latin America, and Southeast Asia). In these regions, the majority of the population is subject to minimum living conditions. On the other hand, there are developed countries where more people live above normal living conditions. Life in such large agglomerates (cities, metropolises) offers its residents much better living conditions, but it creates a number of ecological, sociological, psychological, and other problems whose consequences are felt by all residents, as well as services that are obliged to resolve such problems. Despite the fact that the planet is overpopulated globally, there are entire regions inhabited by fewer people than the resources of that region can withstand (parts of Canada, Australia, and Russia). In each country, uneven population density is a problem that creates a series of environmental problems that will continue into the future;
- Uneven accessibility to vital resources is associated with uneven distribution of resources on the planet, as well as with population density. In some regions, people have at their disposal an abundance of arable land, water, and forests, while other regions are ruled by catastrophic drought and famine due to lack of water and food (arable land). This problem is very difficult to solve, for the simple reason that resources cannot be relocated and made available to people in need;
- Existential problems of world population are another characteristic of modern humankind and affect most of the world's population. It is believed that only one-sixth of the total number of people on the planet have satisfactory living conditions (dwelling place, sufficient

food and access to water, employment, education, and medical treatment). Poverty, unemployment, and hunger are problems of sustainable development that pose great challenges for the world and the ways to solve these problems are uncertain;

- The fact that most conventional energy resources are not renewable emphasizes the global problem. Conventional energy resources (fossil fuels, coal) are extremely limited and their use is intensive. Almost all industry in today's world is based on the use of energy that comes from these resources, and they will inevitably cease to be exploited 1 day. Most experts agree that reserves will suffice until, 2020, but after 2050, huge problems may appear in this field. Only rational energy management and prevention of any huge spending can make the situation easier to handle;
- Renewability of only small energy resources as an alternative to the traditional sources implies that there are resources (wood, energy, and water—if water is used rationally and there is no major drought) that can be renewed. This type of energy supply is limited and meets only minor needs of the population. In order to preserve these resources, responsible actions of authorities and society as a whole are required in terms of their conservation and restoration;
- Limited use of alternative energy resources as a way to meet specific needs. Wind energy, solar, biomass energy, and the like are used in a limited way because these ways of obtaining energy are still rather expensive. It is realistic to expect increased efforts in the future to use alternative energy resources more and develop solutions that are appropriate for more users;
- Uneven distribution of natural energy resources is the problem civilization has always been faced with, and the first civilizations were created in areas generally rich in resources. With the increased needs for energy that emerged in the period of rapid industrialization, most developed countries became forced to purchase large amounts of energy in the market, i.e., from imports. Identification of the position of certain energy supplies leads to gradual and then ever-increasing polarization among the countries that have large amounts of energy resources and the countries dependent on them. With the emergence and intensive use of alternative energy sources, this problem can be alleviated to some extent, but it certainly cannot be completely removed. Uneven distribution of energy resources is

one of the important factors of general instability, and the inability to solve this problem imposes the need to develop more advanced forms of global cooperation;

- Concentration of large energy consumers, as another aspect of global energy imbalance, directs the flow of energy and the problems associated with it. Namely, as in most human activities, in the area of economic growth, which is closely related to the use of energy, there is a relatively sharp division between countries that consume the biggest quantities of the world energy supplies and the countries whose consumption is average or even symbolic. Strong economic development in certain parts of the world (Europe, United States, Canada, Japan, China, and Russia) created the need for directing the largest amount of energy in the direction of these countries and regions;
- Energy as a subject of world trade, a phenomenon caused by the energy dependence, or the inability to achieve economic growth without energy exploitation in various forms. Energy resources that are concentrated in certain parts of the world became an extremely complex and important subject of world trade during the last decades of the 20th century. Some countries are very dependent on the import of energy, while for some countries, energy is the main export product. This arrangement opens up opportunities for a number of discrepancies, inconsistencies, and conditioning.

The need for uniform legislation and harmonious development is of particular importance for sustainable energy development since the energy problem has long crossed local borders and become a problem of general, global interest. In some countries of the world, there are numerous legal regulations and measures that promote sustainable development in general. In a great part of the world, rules have been applied that encourage sustainable energy development, but in this respect, there is no general consensus of all countries, rather there are great differences between individual regions, as well as a wide gap between rich and poor countries (Vanegas Cantanero, 2020).

The problems stated above are the most obvious challenges of survival in the future and are therefore subjects of great interest. Given that the concept of sustainable energy development implies the involvement of all organizations and individuals, it is essential that everyone, depending on the activity, redirects their activities so as to contribute to reducing the intensity of these problems.

Legislation

Global sustainable development strategy is based on the application of a certain number of documents, both binding and nonbinding. Following the selection of sustainable growth as the only development path (UN Conference in Rio de Janeiro, 1992), a number of regulations have been passed in this area, particularly in the most developed countries in the world, which are also the biggest polluters.

Energy management-related resolutions are of a special global interest since energy production and consumption in the world bring about numerous problems, conflicts, and economic crises. Exaggerated use of energy substances leads to energy resources exhaustion and pollution of the atmosphere, which has been the main reason for adopting the Kyoto Convention, Kyoto Protocol, and a number of directives mandatory for the European Union member states (Golušin & Munitlak Ivanović, 2011).

The Kyoto Protocol, together with the UN Framework Convention on Climate Change, is an international agreement adopted in December 1997 in Kyoto, Japan. The main difference between the protocol and the convention is that the convention urges industrially developed countries to reduce emission of greenhouse gases (GHGs), while the protocol makes it mandatory. The European Union is highly dependent on imported energy and, at the same time, a huge emitter of harmful gases and has taken the following measures aimed at finding resolution to the issue:

- Introduction of energy saving and rational use by application of energy efficiency measures, achieving a total savings of the final energy up by 20%–40% through transport, distribution, and consumption;
- Energy saving and rational use measures by application of energy efficiency measures in building, community energy, industrial energy, transport, and other places;
- Replacing fossil fuels with renewable energy sources, achieving lower levels of import dependence, decreased emissions, and environmental protection, including new vacancies.

The Kyoto Protocol came into force on February 16, 2005. By the end of 2008, 183 states ratified the protocol. Its main goal is reduction of global anthropogenic emission of gases with the greenhouse effect by at least 5% in comparison with the reference year of 1990, in the course of the first mandatory period between 2008 and 2012. Annex A of the Protocol requires reduction of six gases with the greenhouse effect (GHG): carbon dioxide (CO_2), methane (CH_4), nitrous oxide (N_2O), and industrial gases, such as hydrofluorocarbon (HFC), Perfluoro isobutene (PFC), and the sulfur hexafluoride (SF6) group.

Annex B contains a list of state signatories and quantified responsibilities pertaining to decreased emission of GHG in percentages in comparison with the reference year. In other words, quantified responsibility of reduced emission of GHG has been envisioned for 38 industrially developed countries, including 11 central and east European countries with transitional economies. In addition to this, the protocol allows states to decide by themselves which of the listed six gases will be included in their national strategy for reduction of gases emission.

It is important to note that the industrialized states with quantified responsibilities listed in Annex B of the Kyoto Protocol are actually developed countries. State signatories that are not in the Annex list do not have the responsibility of a quantified decrease of GHG emission within the Kyoto Protocol. However, they have to fulfill general liabilities envisioned by the convention and the protocol. Other significant directives pertaining to the energy policy of the European Union are as follows.

Directive 2001/77/EC

Directive on promotion of electricity from renewable energy sources in international energy market

Promotion of renewable energy sources (RESs) implies implementation of a number of activities aiming at long-term reinforcing and maintaining of a certain level of energy production from renewable sources, which is carried out by passing adequate development strategies in every country, identifying national goals, and creating a support system and origin guarantees—green certificate, short and simple administrative procedures, defining conditions and tariffs to join the network, etc.

Directive 2009/28/EC

Directive on promotion of the use of biofuels and other renewable fuels for transport

This directive refers to increased use of biofuels in the market by 10% of the total fuel quantity for transport by the end of 2010.

Directive 2001/80/EC

Directive on emission limits in the air from huge combustion systems

This directive refers to huge fueling systems of thermal power higher than 50 MW, for which it stipulates strict emission limits in the air for new and existing power plants that are to be achieved by application of the best available technologies. Its full implementation is supposed to be completed by December, 2017 due to the significant investment required from the energy sector in order to maintain the standards stipulated by this directive.

Directive 1999/32/EC

Directive on reduction of sulfur content in liquid fuels

Implementation of this directive is envisioned to be completed by December 31, 2011 by the Annex to the Energy Community Treaty and pertains to the reduction of emissions and setting limits to sulfur dioxide. In the period of 2000–10, the total emission of sulfur dioxide in the European Union member states went down by 60%, thanks to application of directives on huge systems (European Commission, 2016).

Directive 96/61/EC

Integrated pollution prevention and control (IPPC directive)

In essence, the IPPC directive is about achievement of integrated prevention and supervision of pollution caused by different activities (comprising the whole energy sector). The IPPC directive stipulates measures for the prevention or reduction of emission in water, air, and land, including measures pertaining to waste, aimed at reaching a high level of environmental protection.

This Directive is highly important for the energy sector since thermal energy plants with heating input above 50 MW and oil, mineral oil, and gas refineries need to obtain an integrated permit for the future work of such systems. The year, 2015 is the deadline for obtaining the integrated permit issued by authorities of existing plants and activities. The permit stipulates conditions guaranteeing a high level of environmental protection as a whole.

Case study

From Kyoto to Paris: Measuring renewable energy policy regimes in Argentina, Brazil, Canada, Mexico and the United States, Energy Research & Social Science, Volume 50, 2019, Pages 82–91. https://www.sciencedirect.com/science/article/abs/pii/S2214629618301245.

References

Agenda 21, United Nations conference on environment & development Rio de Janeiro, Brazil, 3 to 14 June. (1992).

Andrews, R. N. L. (2020). *Managing the environment, managing ourselves: A history of American environmental policy.* New Haven, CT: Yale University Press. https://doi.org/10.2307/j.ctvwrm3ww.

Brundtland commission report: Our common future (Oxford University Press). (1987).

Delina, L., & Janetos, A. (2018). Cosmopolitan, dynamic, and contested energy futures: Navigating the pluralities and polarities in the energy systems of tomorrow. *Energy Research & Social Science, 35*, 1–10.

Environmental consequences of the Chernobyl accident and their remediation: Twenty years of experience, International Atomic Energy Agency, Vienna. (2006).

European Commission. (2016). Directive (EU) 2016/802 of the European Parliament and of the Council of 11 May 2016 relating to a reduction in the sulphur content of certain liquid fuels.

Filipović, S., Radovanović, M., & Golušin, V. (2018). Macroeconomic and political aspects of energy security. *Renewable and Sustainable Energy Reviews, 98*, 428–435.

Golušin, M., & Munitlak Ivanović, O. (2011). Kyoto protocol implementation in Serbia as precognition of sustainable energetic and economic development. *Energy Policy, 39*(5), 2800–2807.

Gunnarsdottir, I., Davidsdottir, B., Worrell, E., & Sigrugeirsdottir, S. (2020). Review of indicators for sustainable energy development. *Renewable and Renewable Energy Reviews, 133*, 11294.

Hicks, J. R., & Simon, H. (1979). *Modes of economic behavior: Variations on themes.* Cambridge, MA: Massachusetts Institute of Technology, Department of Economics.

Ing, J. (2020). Adverse selection, commitment and exhaustible resource taxation. *Resource and Energy Economics, 61*, 101161.

Jha-Thakur, U., & Khosravi, F. (2021). Beyond 25 years of EIA in India: Retrospection and way forward. *Environmental Impact Assessment Review, 87*, 106533.

Lacey-Barnacle, M., Robison, R., & Foulds, C. (2020). Energy justice in the developing world: A review of theoretical frameworks, key research themes and policy implications. *Energy for Sustainable Development, 55*, 122–138.

Lior, N., & Radovanović, M. (2017). Sustainable economic-environmental planning in South-East Europe—Beyond-GDP and climate change emphases. *Sustainable Development, 25*, 580–594.

Lior, N., Radovanović, M., & Filipović, S. (2018). Comparing sustainable development measurement based on different priorities: Sustainable development goals, economics, and human well-being—Southeast Europe case. *Sustainability Science, 13*, 973–1000.

Luna-Nemecio, J., Tobón, S., & Juárez-Hernández, L. G. (2020). Sustainability-based on socioformation and complex thought or sustainable social development. *Resources, Environment and Sustainability, 2*, 100007.

Mallory, M. L., & Ando, A. W. (2014). Implementing efficient conservation portfolio design. *Resource and Energy Economics, 38*, 1–18.

Matthews, J. C., Allouche, E. N., & Sterling, R. L. (2015). Social cost impact assessment of pipeline infrastructure projects. *Environmental Impact Assessment Review, 50*, 196–202.

Meadows, D. H. (1974). *The limits to growth: A report for the Club of Rome's project on the predicament of mankind.* Universe Books, ISBN:978-0876631652.

Mirković-Isaeva, O., Radovanović, M., & Vasić, B. (2018). Natural resources curse and geopolitical challenges of green economy. In *International scientific conference "green economy and environmental protections", proceedings, pp. 116, 23-25 April, Belgrade, Serbia,* ISBN:978-86-89061-11-6.

Primc, K., & Slabe-Erker, R. (2020). Social policy or energy policy? Time to reconsider energy poverty policies. *Energy for Sustainable Development*, *55*, 32–36.
Repetto, R. (1988). *The forest for the trees?*. Washington, DC: World Resource Institute.
Report of the world commission on environment and development: Our common future, presented to the UN general assembly in 1987. (1987).
Roche, P. K., & Campagne, C. S. (2017). From ecosystem integrity to ecosystem condition: A continuity of concepts supporting different aspects of ecosystem sustainability. *Current Opinion in Environmental Sustainability*, *29*, 63–68.
Sauvé, S., Bernard, S., & Sloan, P. (2016). Environmental sciences, sustainable development and circular economy: Alternative concepts for trans-disciplinary research. *Environment and Development*, *17*, 48–56.
Scoones, I., Stirling, A., Abrol, D., Atela, J., Charli-Joseph, L., Eakin, H., et al. (2020). Transformations to sustainability: Combining structural, systemic and enabling approaches. *Current Opinion in Environmental Sustainability*, *42*, 65–75.
Solow, R. M. (1993). Sustainability: An economist's perspective. In R. M. Dorfman, & N. S. Dorfman (Eds.), *Economics of the environment—Selected readings* Norton.
Vanegas Cantanero, M. M. (2020). Of renewable energy, energy democracy, and sustainable development: A roadmap to accelerate the energy transition in developing countries. *Energy Research & Social Science*, *70*, 101716.

CHAPTER 3

Strategies for sustainable energy management

Contents

Sustainable Energy Management
https://doi.org/10.1016/B978-0-12-821086-4.00007-6

Basic approaches to implementing strategy of sustainable energy management

The implementation of the process of sustainable energy management can be undertaken by using appropriate techniques that are applicable no matter whether the management process will be applied in a particular company, region, or a country. Application of these methods of energy management is not recommended on a level broader than national, since the process requires analysis of the environment, and the control involves measuring performance in each state, so the accuracy is acceptable up to this level. A particularly unclear situation will result from the analysis of an environment that is very complicated in an area wider than one country, as among countries there are significant differences in terms of social, political, economic, and environmental situations (Armstrong, 2001).

The main role of the methods discussed in this chapter is to implement the previously selected strategy of sustainable management of energy. The strategy of sustainable management is the only acceptable strategy for energy management, and its implementation can be carried out by selecting one of the proposed models. All of the following models are designed to allow segmentation of the subject on which the control is performed. The company, region, or country in question is characterized by its specific characteristics, determined by many parameters, and in the given time, it undergoes various stages of development.

Traditional management approached energy solely on the basis of its cost-effectiveness. Modern sustainable management imposes the need to analyze various aspects of energy efficiency, consumption of energy, and harmful emissions that result from it (Thakkar, Chaudhari, & Shah, 2021). Cost-effectiveness is no longer the only factor that influences the decision-making process. The environmental component, and energy sustainability within it, is becoming a very important factor that to some extent changes the traditional concepts of management and absolutely disrupts the traditional primacy of cost-effectiveness.

The economic aspect of management is still the subject of study, but by no means is it the only and the most important aspect. The purpose of the new analysis of management techniques serves to find a method that harmonizes economic growth and environmental concerns—energy responsible business operations—without substantial harm to the development of a country or company as a whole. In current practice, in order to implement sustainable management of energy, six basic approaches have been defined:

- Traditional approach;
- System approach;
- Eco-management approach;
- Total quality management approach;
- Life cycle analysis approach; and
- Gap analysis approach.

It cannot be said which of these approaches is the best, because each has its advantages and disadvantages and specific features that make it more or less acceptable under certain conditions. It is often desirable to combine these approaches to get the best results.

Traditional approach

The traditional approach is a logical way to implement the sustainable management of energy; its basic model is shown in Fig. 3.1.

Development of program

The traditional approach is a "top-down" approach. This approach begins with obtaining management support and determining program goals before collecting necessary information. The aim of the first step is to provide an executive-level management decision to start developing the strategy of sustainable management of energy that must be incorporated into the business policy of the company (Goel, 2007).

At this stage, a team is created to collect necessary data, inspect the areas of application, and set priorities for implementation of the program. A preliminary evaluation is necessary to obtain the information needed to

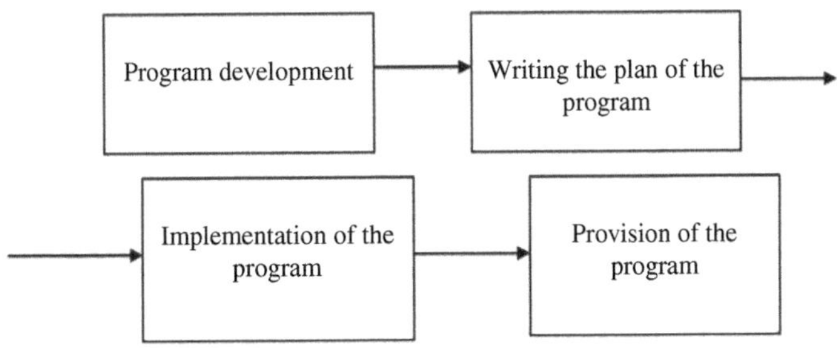

Fig. 3.1 The traditional approach to implementing the strategy of sustainable management of energy.

create the initial strategic plan. This part of the strategy is carried out by a team of employees or a group of external experts.

At this stage, the team collects information that might be potentially important: the amount of energy consumed, energy sources, owned energy resources, energy consumption per workplace, per process, and per product, and the level of energy efficiency and energy waste, as well as economic aspects of energy consumption. This information should provide a realistic picture of the current state of energy management and therefore, identify the shortcomings and problems that will enable clear insight into the priorities of the energy policy.

Writing a plan of the program

The second step of the traditional approach is to write a program plan for the sustainable management of energy. A good plan should indicate what all stakeholders want from the program, including consumers, suppliers, employees, legislation, environmental organizations, local communities, shareholders, and others who have any interest in the program implementation.

The plan should indicate the goals clearly and the ways to reach them. It is not enough to decide to increase energy efficiency. It is necessary to clearly define the level of energy efficiency that is to be achieved and indicate the means and resources necessary to achieve that goal. Possible obstacles need to be anticipated in terms of implementation, and the plan should be made so as to overcome the obstacles. Finally, the plan should determine the schedule of activities that will, among other things, enable management to monitor the process and progress during the year (Huffmire & Holmes, 2006).

Implementation of the program

Implementation of the program of sustainable management of energy by the traditional approach can be summarized, as shown in Fig. 3.2.

The aim of the detailed evaluation is to help the team determine alternatives, and for this, *brainstorming* is used. The traditional approach does not include a formal analysis of the root of the cause before the alternatives have been determined. This phase involves appointing the team for evaluation, data review and organization, and documenting of information. It is necessary to precisely determine the existing gaps in energy policy, which includes estimation of the type, intensity, and durability of problems. Detected shortages are documented and ranked according to their urgency (Bonnardel & Didier, 2020).

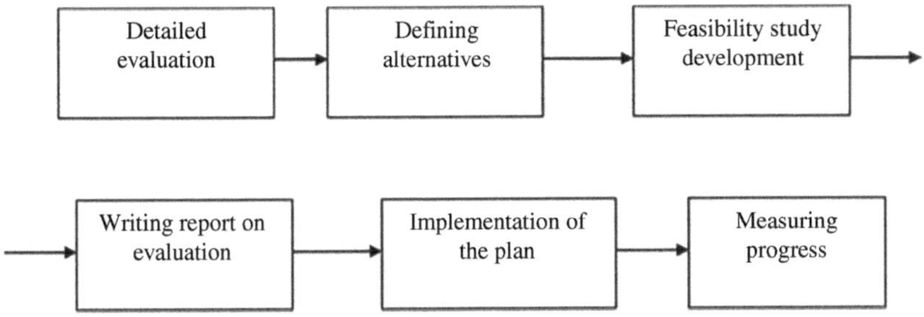

Fig. 3.2 Traditional implementation of a program of sustainable management of energy.

Based on the detailed assessments of the possibilities for resolution of the identified problems, the evaluation team proposes a certain number of alternatives and screens to help determine the focus of the implementation that follows. Generally, a limited number of alternatives are recommended. The traditional approach for the assessment of alternatives uses a matrix of criteria. For defining the problem to be solved, the following criteria are often used: degree of urgency, technical and technological limitations, availability, capacities, and economic aspects.

Upon completion of the review, the analysis of the feasibility of some alternatives is started. Of course, it is not necessary to conduct a detailed analysis for each alternative respectively, but on the other hand, there are those that are particularly demanding. Feasibility analysis should indicate the technical, environmental, and economic feasibility of each alternative. In most cases, it is first necessary to consider the technical feasibility of the proposed solution model. In fact, some energy problems could be resolved in a certain way, but usually there are technical limitations that hinder the application in some way. It may be that it is necessary to use a more expensive solution, because the cheaper and more accessible ones may not be technically feasible.

After evaluation of the feasibility of the alternatives, a report should be written indicating the details of the analysis and pointing to information that should be presented to management. In addition to technical feasibility of the chosen solution, it is necessary to point out the specific benefits of the proposed improvements, such as the amount of energy that can be saved, the degree of increasing energy efficiency, reducing waste, the amount of energy gained from renewable sources, etc. In this way, management can estimate the positive financial and nonfinancial effects of the proposed change.

Upon approval by management, the next stage entails implementation of the plan, which includes project selection, engaging necessary resources, and starting implementation of alternatives (McGlynn, McGurk, Effland, Johll, & Harding, 2004). The sustainable management of energy is approached through prior achievement of the strategic goals that were selected based on the previously defined criteria. These criteria create the traditional method of management, which is closely problem-oriented because it responds only to the perceived problems and thus has a predominantly reactive nature.

During the stages, the whole process is monitored and adjusted in order to achieve the set goals. The traditional approach does not require preparation of a specific action plan. The last step of this phase is to measure progress. The traditional approach requires the collection and analysis of the data, as well as measurement of economic results.

Provision of program

In these stages, five activities are performed, as shown in Fig. 3.3.

Provision of the program begins with the integration of the program of sustainable management of energy in formal corporate initiatives and plans. These plans may include safety, quality, preventive measures, and the like.

Education and training of people involved in the program must be specified. Participants must be familiar with the process, especially if there are new employees. Employees must undergo an annual training and testing of knowledge about the program and constantly follow up on innovations (Krampell, Solís-Marcos, & Hjälmdahl, 2020).

Communication is important for the implementation of each activity. The traditional approach seeks to develop all types of communication. Communication should be two-way, and it is necessary to consider the suggestions of employees. The traditional approach recommends particular consideration of the program of rewarding employees.

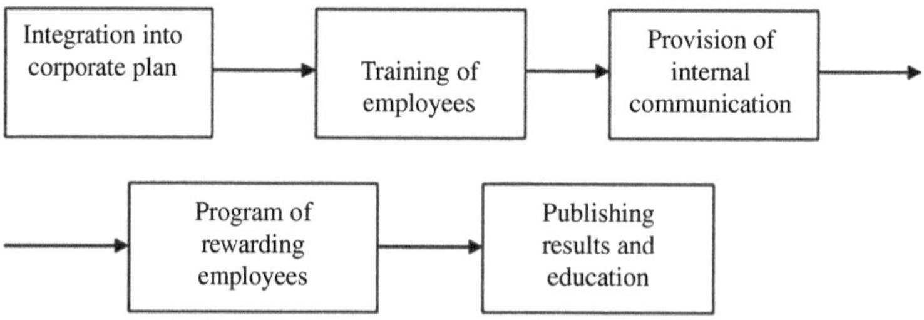

Fig. 3.3 Provision of program.

System approach

The organization is a system if it operates as a whole and if its parts develop certain interactions. The system approach to sustainable energy management observes the whole organization, as well as its parts and connections among these parts. The functionality of the organizational units largely depends on how they are connected, more than on the parts themselves (Whitehurst & Hamel, 2015). This model envisions that the effectiveness of the sustainable energy management system is more affected by the way and efficiency of the connections that exist among organizational units than by the specific features of the parts. However, the system approach is based on the fact that energy is not lost but is transformed from one form to another (Antony & Govindarajan, 2017). According to this model, each system on which sustainable management of energy should be implemented is made of a set of components and connections between them, which are characterized by certain flows of energy. Parts of the system are connected directly or indirectly, and the change in one part affects all other parts. The system approach is unique for several important reasons.

Processes that consume energy do not always provide information that is easy to present. Instead, these processes are sometimes defined so that the amount of energy consumed per process, per product, or per work task cannot be determined. For this reason, too much time and energy are spent in trying to understand the situation before moving on to the next stage of solving the problem (Ziębik & Gładysz, 2018).

The system approach cuts across this very situation. System tools indicate how the situation can change and how the resources can be saved. This is the basic difference between the application of the system tools and engagement of an external expert for energy management while the employees are not involved in the process. The system approach involves decision-making by teams within the organization, rather than individual experts. The team members are required to analyze the problem of energy management to determine the cause of the problem and offer alternatives for solutions. That does not mean that teams should not use the services of experts. However, the responsibility for making decisions should be given to the team members who will go further in the implementation and evaluation (Namany, Al-Ansari, & Govindan, 2019).

Since the system approach is interactive, team members often believe they are responsible for making all decisions. Of course, other employees have their own opinions. Another problem is the very process of management in the organization. It is important to identify the reasons for the use of resources and for losses before arguments have been provided so managers can make decisions about the program. In this context, the system approach provides the management a clear view of processes and problems. Only in this way can management effectively plan strategies and objectives.

Many organizations believe that adjustment is the basis for survival in the global economy. Managers learn new ways to run their organizations and employees learn how to contribute to the improvement of business by using their knowledge and skills. By learning how to supervise, control, and continuously improve production, organizations are able to meet the demands of consumers and other stakeholders. The principle of qualitative improvement can be a useful tool for achieving environmental excellence. Application of tools for improving quality is a powerful means to eliminating environmental inefficiency and preventing pollution.

Stages of the system approach

The system approach to implementation can be presented in several ways, depending on the degree of details, but is generally as described in Fig. 3.4.

Mapping the process. Learning more about the use and waste of energy resources during the process of work and updating the existing knowledge are the two main features of this process. These tasks require the collection, reviewing, sorting, and comparison of information. The characterization of the process is a step during which most knowledge about energy cycle in the company is acquired. At this point, the existing knowledge of the process is reviewed and reshaped, and new knowledge is easily acquired and complements the existing information, because the procedure of process mapping allows the process to become "visible." It is now possible to observe the

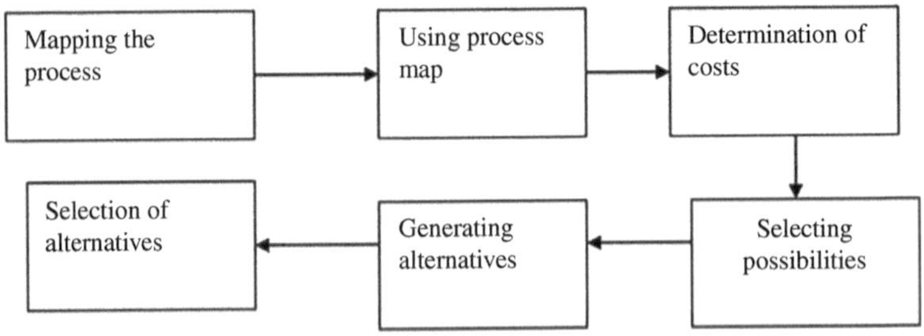

Fig. 3.4 Stages of the system approach to implementation of the strategy of sustainable management of energy.

process in a new way and look at all operations for support, in order to identify how they affect the progress of the main process.

Characterization of the process of flows and consumption of energy is more efficient than preparation of the classic worksheets. Diagrams are always a better and clearer way to represent information. Links that are identified between the process steps help to identify more clearly the causes and places of the use of resources and waste generation. Every place resource is used is an opportunity to save those resources. Each loss of energy resources is an opportunity to avoid that loss (Zhang, Wei, & Zhou, 2018).

Characterization of the process includes collection of all information that describes the activity that is under observation. An effective way to create the characterization of the process is the use of the so-called hierarchical process maps. In most organizations, process documentation is created under categories such as company, equipment, production line, or department. The information is given in the form of diagrams, maps, machine configuration, plans, or in some other way. Each of these methods of characterization of the process has a fundamental flaw due to its great complexity because there are too many objects on a single sheet of paper.

Research has shown that most people can simultaneously observe six objects and express what the objects mean and the connections between them (Smith et al., 2019). Hierarchical process characterization shows a maximum of three to six objects on a single sheet of paper, because almost every process can be shown in such a way. Subprocesses are also shown at the next level, but again, displaying only three to six objects. In this way, a sort of "tree" is constructed, which shows the relationship between process steps. The system approach indicates that each step is connected with the other steps and that they all create the system as a whole. The hierarchical process structure can be summarized as shown in Fig. 3.5.

During the mapping process, two basic rules need to be observed. First, the process map should help the program implementation team understand the process better than using other methods. Second, process maps should help the team communicate better with management and other stakeholders.

Using process maps as patterns. Some organizations think about the process as a box with inputs and outputs. The use of this method is appropriate for introducing only limited changes and enables the application of the concept of sustainable energy management. Using process maps as patterns, process documentation can be organized into separate steps in the process at the least number of levels. All procedures for the implementation of standards, the best management practices, implementation of recommendations, and

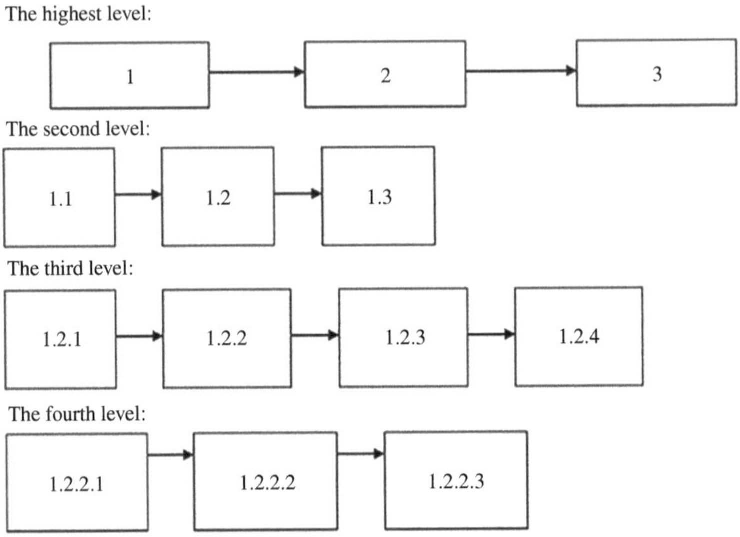

Fig. 3.5 Hierarchical structure of the process map.

others can be presented through the steps in the process map. What can be noted when using the process maps is that many problems are associated with a single step, and this makes it easier to focus activities on that particular step (Fleacă & Fleacă, 2016).

To present the process maps, many designers use simple pencil and paper, although they can be presented using a computer and special software. If the organization decides to computerize process information, everyone involved in a single working step should have access to all information. The use of process maps as patterns helps organizations track the use of resources and waste in each process step in the main or side processes. All resources (energy, water, raw materials) can also be followed up through some process map. A simple review is shown in Fig. 3.6.

The term "nonproduction use" means that the observed energy resource does not become an integral part of a semiproduct or the final product. The term "nonproduction loss" means that the energy resource is lost in that step as waste, discharge, or emission into the environment. Process losses can be classified on the basis of the medium (air, water, solid waste). Costs can also be monitored on the basis of the process steps. Monitoring of this information is useful to rank the changes according to costs. All this can be provided by application of software, as well (Waqas Iqbal & Kang, 2020).

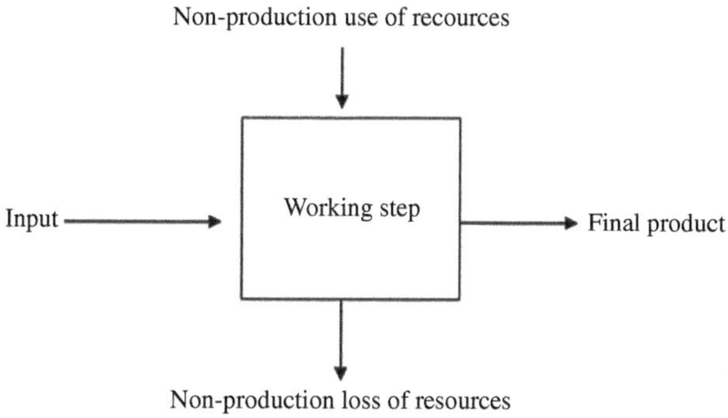

Fig. 3.6 The use of a process map as a pattern for following energy resources.

Determining the cost of losses. Collecting accurate data on costs is important to justify the investment in alternative solutions. These costs are collected by the operating steps because these are the places where the program of sustainable energy management will be implemented. There are three types of costs to be monitored: total costs, costs incurred by the loss of energy resources, and costs based on the cost of management and other intangible costs.

For each loss identified in the process map, the program should examine the so-called calculation map in order to check whether the cost is tracked by accounting services. The costs map shows a number of costs and/or a different code for this category of payment. It is important to remember that usually only the money that comes into or goes out of the organization is tracked. Internal transactions are not tracked. The internal transactions are the costs of activities. All data on costs obtained in this way are accurate in most cases. The existing procedures must be applied to energy costs, too (Haraguchi, Siddiqui, & Narayanamurti, 2019).

The second category of costs is associated with the costs of resources that are nonproduction output or process losses. For example, painting of new cars in the car industry is connected with the use of sprays. Some cars will not be completely covered in paint, because a certain amount of spray is retained on the filter or ventilation system. It can happen that 60% of the parts are well painted, while 40% are not. In this case, the whole process of painting should be conducted again. The 40% is the loss of that working step, and that is a nonproduction loss. The costs of wasted color and energy can be added to the total cost for color filters, which increases energy nonproduction costs.

The third category of costs is associated with the losses regarding the activities of management. If the costs are covered (e.g., hazardous air pollution, hazardous waste, priority waste water), there are a number of activities that are regulated. First, activities have to be determined that occur for each nonproduction cost for each working step starting from the lowest level of the process map. Then, costs associated with each of the selected activities have to be estimated.

Often the costs of nonproduction losses are triple if added to the lost resources. If the losses are regulated, the costs of activities of managing the losses can increase the costs up to five times. The effort should be focused on the registration of all costs of losses to prevent them and save money.

Selection of opportunities. The information gathered during the stage of mapping the process can be used in order to select the opportunities that should be focused on in the process of problem-solving and decision-making. Some programs select the opportunities by trying to eliminate problems associated with the use or loss of energy resources. Other programs seek to identify opportunities among those that were previously selected by managers or specialists in the field of energy. Each organization has its own method for selecting opportunities, and there is a special tool (which dates back to 1897) that can help to easily perform this operation, discussed in the following (Haley, Gaede, Winfield, & Love, 2020).

If all opportunities identified in the process map are ranked according to the sequence of their costs, it can be noted that 20% of them carry about 80% of all costs, and vice versa. In most cases, this 80/20 rule (the so-called Pareto principle) should be used for selecting opportunities. Most organizations apply this rule in their work (Horowitz & Woo, 2006).

Quality improvement experts recommend focusing on several vital sources of the problem, thus avoiding their interference with the issues of minor importance. The term for this process is called ranking. Pareto analysis is a tool for implementation of the ranking. Potential opportunities should not be rejected easily, although in the early stage of implementation of sustainable energy management, a careful selection has to be made so as to choose the most likely chance of success. Most often, teams have to work with projects that are either very large or very difficult to handle. We must focus attention on the 20% of causes because they bring about long-term benefits. In the process of sustainable energy management, it always happens that in 20% of the places in the organization, 80% of problems have been identified regarding management of energy. Thus, attention should be directed toward the possibility of their elimination.

Cause analysis. The cause is the main reason why the energy resource is used or the loss occurs. If this cause cannot be eliminated, the use or the loss of the energy resource should be prevented.

Analysis of the root of the cause refers to identification of causal factors. Most people involved in the program of sustainable management of energy are committed to resolving the problem, but in their attempt to find a solution, they may quickly overlook some very important activities in the process of problem-solving. A team that skips important steps will often take the obvious actions instead of those that provide the best possible solution.

For example, when faced with environmental problems caused by excessive use of fossil fuels, the members of the problem-solving team may assume from the start that the best solution to the problem is to find an adequate and safe substitute—energy from alternative sources. In fact, the problem may be caused by the way in which the fossil fuels are used in the organization and not by the features of the fuels themselves. The change in procedure or equipment or staff training is often the best solution for reducing costs of this kind. Root cause analysis teaches organizations to consider all potential causes: material, technology, work processes, and employees.

Root cause analysis can be an effective management tool to determine the real cause of the problem, the ways to modify equipment, etc., and the preventive measures needed. It also provides opportunities for improvement (Bhattacharjee, Dash, & Oaul, 2020).

A cause-and-effect diagram, also known as a fishbone diagram, is the most commonly used tool for solving these types of problems. To be used effectively, it requires some training and experience and should be used by the whole team. This diagram clearly indicates to management and other stakeholders what may be the cause of the problem in energy management. Once a diagram has been made, the team members need to accurately count the causes that have been identified (Luo, Wu, & Duan, 2018). The 80/20 rule can help to focus attention on the most likely cause by drawing circles around the 20% of causes that result in 80% of problems. The team will be more effective if they focus on the process of identifying problems before attempting to propose solutions.

Generating alternative solutions. Each approach has a certain method for generating alternatives to solving problems in energy management. Some experts believe that practical applications should be limited to a small number of alternatives for solving a specific problem, either because they have not performed the analysis of the root of the problem or because the team members are not adequately involved in the process of generating alternatives. References and practical examples provide just a few ways to specify each problem. The team members in each organization should be satisfied if they can develop even a few alternatives.

Tools for improving implementation of the system approach

Tools have been developed in order to make the system approach easier to use and more efficient. The two tools that are most commonly used are action plans and checklists.

An action plan should be prepared for each activity that is planned for each year of implementation of the program of sustainable energy management. Regardless of how the organization plans to implement certain new solutions, the stage of formalizing its own action plan should not be neglected. Each action plan should show all the alternatives that will be applied and indicate substages and steps that will take place. The action plan should indicate the name of the person responsible and accountable for the performance of each step (Amorim, 2014).

Each step should lead to achieving a recognizable goal that should also be clearly indicated. It is necessary to mark the measurements necessary for monitoring the progress during implementation in order to reach the defined goal and determine the time frame for completing each step. Finally, it is necessary to determine the resources for realization of the whole process. An example action plan is given in Fig. 3.7.

Checklists are another helpful tool for the sustainable management of energy. Checklists help conduct organizational activities and progress and provide important information on the steps and methods for measuring organizational performance and effectiveness. They also help organizations collect and organize data necessary for evaluation of the current status and for monitoring the process. It is useful to compile a list of questions and answers for everything associated with each tool for problem-solving and decision-making.

Selected alternative:			Date:		
Activity	Person responsible	Performance	Monitoring technique	Deadline	Required resources
1.					
2.					
3.					
4.					

Fig. 3.7 Example action plan.

Checklists help the team properly conduct future activities and suggest which things should be done, the people with whom to contact, and what questions to ask. Checklists also help the team organize tasks and gain general insight into the situation, recommendations, characteristics, alternatives, and consequences. The following steps should be taken to create a quality checklist (Tsani & Menegaki, 2018):

- Determine the purpose of using the checklist.
- Conduct research to ensure the checklist covers all recommendations and answers to questions that should be collected.
- Provide space for checking steps that are finished.
- Ask experts in the concerned area to evaluate the final draft of the checklist to ensure all important details are included.
- Provide audit and pilot testing of the checklist prior to its application in practice.

Teams should create a checklist that follows the process envisioned by their program. Some components of typical efforts use system tools, so teams can use a checklist to:

- Determine the real costs for each working step in the process;
- Collect information necessary for ranking alternatives;
- Select alternatives that will be analyzed;
- Ensure that all causes of the problem are taken into consideration;
- Document the search for potential solutions and alternatives;
- Collect information on each alternative in order to determine priorities;
- Test complete action plans;
- Track information during the process;
- Test the use of each element in the process; and
- Test the total efficiency of the program.

Periodic use of checklists creates the impression of continuous assessment of progress. Checklists should be designed to provide managers and teams a starting point for the assessment of important characteristics of each step in the system approach, giving answers to questions of what to do and analyzing whether the tools are properly used.

Combining the traditional and system approaches

Most managers use the traditional approach for the implementation of sustainable management of energy. The recommendation is to start by considering changes offered by the system approach that will improve business.

The area that can be particularly improved in this way is the characterization of the process. The system approach requires development of process maps that allow people to easily identify and get insight into areas on which they should focus. The system approach is particularly effective for monitoring the flow of energy.

Mapping of the process should be done before implementation of the program is proposed because it will be easier to point out the reason for the change. Another potential improvement offered by the system approach is using problem cause analysis in order to select options. Experience has shown that teams that do not use problem cause analysis are able to identify no more than two or three alternatives. Teams that use the root of the cause analysis indicate that there as many as 40 alternative solutions.

The traditional approach emphasizes setting the objectives first in any business. The system approach determines objectives in the action plan but only after the information has been collected and analyzed. The goals are changed according to the needs and vary from one project to another, from 1 year to another. Some experts believe that in terms of quality, the biggest problem of the traditional approach is inadequate setting of goals.

Eco-management approach

The latest series of international voluntary standards known as the ISO 14001 series is an effective tool for improving organizational environmental performance and implementation of a sustainable approach to energy management. The purpose of the standards is to establish and implement a systematic management plan that is designed to continually identify and reduce environmental impacts arising as a result of organizational activities, products, and services (Johnstone & Hallberg, 2020).

The eco-management approach may be particularly suitable because its implementation is a possible way to replace the widely used environmental control system based on legislation and controlling of the application of environmental and energy regulations. Eco-management systems can help organizations integrate environmental efforts into everyday efforts, decision-making, and business practices.

Application of an eco-management system improves compliance with environmental regulations and prevents organizations from making progress without taking into consideration environmental regulations so that in both cases it has positive effects on reducing negative environmental impacts. Today, the number of companies in the world that integrate the eco-management system in their business strategies is rapidly increasing. They accept advanced technologies and achieve some savings. Systems of eco-management encourage companies throughout the world to consider environmental consequences of their operations and define strategies to help them reduce waste, risks, and costs (Orcos, Pérez-Aradros, & Blind, 2018).

As with other management systems, eco-management is a formal approach to setting goals, decision-making, information gathering, measuring progress, and improving business performance. Eco-management promotes important elements of planning and improvement required to make an impact on all forms of environmental pollution. In addition, certain elements of eco-management provide an opportunity for implementation of assessment and continuous evaluation by management. A simplified description of the implementation of a sustainable energy management system using eco-management is given in Fig. 3.8.

The eco-management approach is based on a documented policy that includes three key principles: compliance with changing environmental regulations, pollution prevention, and continuous improvement of environmental performance (McGuire, 2014). In some cases, corporate policy can be complex and general and as such unclear to employees and the public. Therefore, the eco-management policy should focus only on the three basic principles given above.

The eco-management system identifies and adapts environmental, legal, and voluntary standards related to the work of employees and business partners. Voluntary recommendations include a pollution prevention program, health and safety prevention, and sustainable development. In addition, eco-management approach for implementation of energy management elaborates procedures on how to achieve objectives and also determines environmental responsibility and ways to respond to requests of individuals, organizations, and other interested parties.

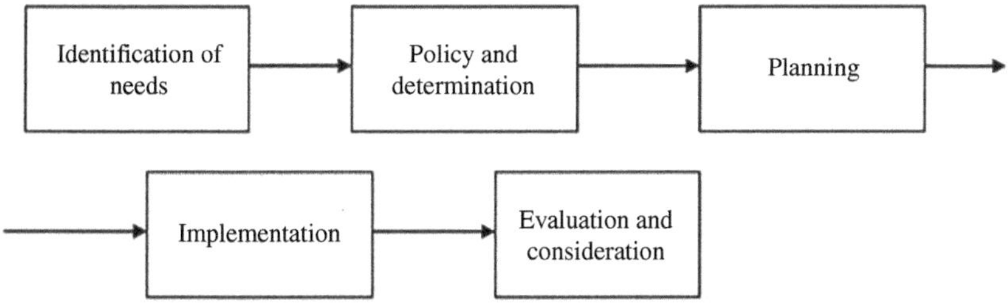

Fig. 3.8 Implementation of a program of sustainable management of energy using the eco-management approach.

Identification of all aspects and determination of their significance are usually the biggest problem in most companies. Because of that, the eco-management system determines procedures necessary for identification of environmental impacts of business activities, products, and services. Thus, most organizations are focused almost exclusively on their negative environmental impacts. Positive environmental impacts are also important, such as recycling and hazardous waste collection sponsored by the company (Martins & Fonseca, 2018).

The eco-management system establishes specific goals to implement energy management initiatives and ensures the participation of employees, as well as technical and financial support. In setting goals for environmental improvement of each business function in the company, the eco-management system recognizes existing environmental laws, recommendations, and requirements of all stakeholders (Comoglio & Botta, 2012). The eco-management system is based on documented procedures for the prevention, detection, investigation, correction, and reporting (internal and external) of active or potential environmental impacts and emergency situations. If the environmental incident occurs, the eco-management system should be reviewed immediately and adjusted so as not to allow such an incident to repeat. The eco-management system also provides for adequate training of employees, suppliers, and other business partners. Organizations should review the competencies and capabilities of its employees to meet the requirements of implementation of the eco-management system.

A particularly important factor in the successful implementation of the eco-management system is the determination of top management, who should accept that the eco-management system is more adequate than other management systems. With this regard, top management should provide maximum efforts to implement eco-management as well as review it periodically to improve it. Over time, investments in the implementation of the eco-management system prove justified, due to the improvement in environmental performance of the business, which satisfies all the parties involved.

Identification of needs

The aim of the ISO 14001 standard is to establish a common approach to eco-management that is internationally recognized. The ISO 14001 standard is a standard for management and not a standard for performance. Eco-management provides the system approach for the integration of

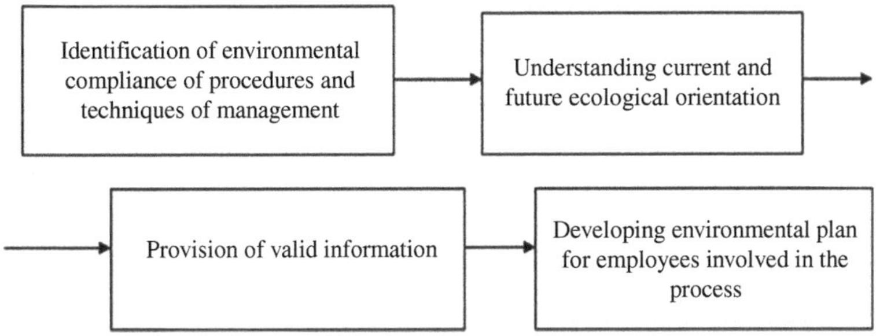

Fig. 3.9 Identification of needs.

environmental protection in business functions and management strategies, primarily because it requires developing appropriate business policies by top management. The eco-management approach involves several initial activities that take place in the sequence of events given in Fig. 3.9.

Many organizations find it useful to use gap analysis of the existing energy system as an initial step in the development of an eco-management system. In this way, the organization is able to review the current status of its own energy performance and identify areas that require attention when applying the eco-management system. (Min, Morgenstern, & Marjanovic-Halburd, 2016) Gap analysis results provide a kind of benchmark regarding compliance of organizational performance with ISO 14001. The gap analysis should include areas of business related to ecological systems as well as links between specific equipment.

The eco-management approach to implementation of energy management encourages organizations to coordinate their activities with relevant environmental legislation and regulations along with other recommendations. The eco-management system together with the ISO 14001 recommendations offers good potential for achieving environmental efficiency and cost reduction. In this way, top management in the company observes the eco-management system as a way to achieve competitive advantage and not as a place for incurring costs. Inclusion of top management in the procedures of policy development, considering the existing plan and ensuring the application of eco-management, is a positive step (Murmura, Libertadore, Bravi, & Casolani, 2018).

Energy policy and determination

Management is responsible for developing the vision and direction of development of the energy system in accordance with the environmental assessment of performance. When defining the energy policy, management should define the scope of observation and ensure consistency with organizational vision, core values, goals, and other efforts (Locmelis, Blumberga, Bariss, & Blumberga, 2017). The energy policy of the company should include a plan to implement the pollution prevention program, harmonization of business operations with existing recommendations, and continuous improvements of the system. The process of defining an energy policy can be summarized as given in Fig. 3.10.

An energy policy is used as a specific guide for developing and understanding organizational energy goals. The eco-management approach to implementation of energy management does not require defining specific energy goals. Instead, it provides a general network for organization and implementation of goals that are necessary for improving energy performance.

After defining the energy policy, it must be documented, updated, and delivered to employees. Many organizations have already developed procedures for internal and external communication of their own policies (Zhou & Dai, 2020). The eco-management approach to implementation of energy management recommends announcing the energy policy of the company publicly. Many organizations already provide certain environmental and energy information in annual reports and plans for dealing with emergencies.

The eco-management approach to implementation of energy management defines the procedure for responding to external demands for environmental and energy information by which it also determines who is responsible for contacts, who notes the date and the nature of the request and response, and whether and what kind of written material is sent in reply.

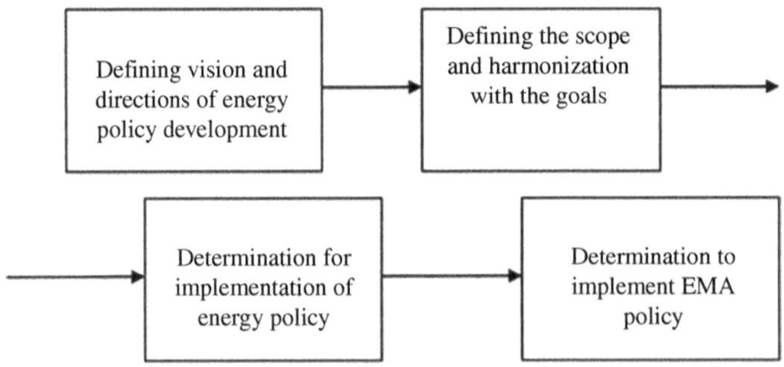

Fig. 3.10 Process of defining energy policy.

The eco-management approach to implementation of energy management also recommends that organizations develop and implement procedures for the purpose of internal communication. The eco-management system and environmental aspects should be associated with all levels of organization and business functions that may have an impact on the environment. The procedure of internal communication outlines the responsibilities in the chain of communication developed for eco-management and environmental issues. Changes may include environmental information, such as review of the objectives, changes in procedures, environmental incidents, etc.

Planning

Planning is needed as the organization grows and production lines change. Planning should go along with other business planning efforts. The stage of planning within sustainable energy management can be summarized as shown in Fig. 3.11.

Insufficiently focused or badly managed efforts will inevitably lead to poor implementation of energy performance and higher costs. On the other hand, well-developed and effectively implemented programs lead to the achievement of high performance and cost reduction. Improved energy performance is an important benefit for most organizations that use the eco-management system for development. Also, some organizations operate under an eco-management system that systematically follows up on environmentally relevant activities. This eco-management system covers the organizational structure, procedures, processes, and resources needed to implement effective environmental management.

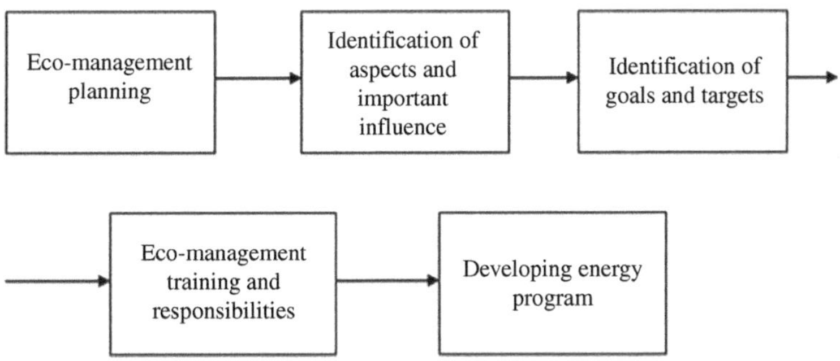

Fig. 3.11 Eco-management planning.

Top management appoints the company managers who will monitor the achievement of the objectives set out in the eco-management system. Managers in charge monitor and evaluate the system and report to top management about eco-management activities. Coordinators work with lower organizational levels and try to generate new ideas that may lead to modification of the eco-management system in terms of its improvement.

Teams are used to multiply the force of the organization. The team approach allows discussion and comparison of different viewpoints and opinions. It may be useful to form independent teams whose members will be selected from all levels of the organization. Involvement of people from different levels of business can only improve the overall efficiency of these issues. Clearly identified teams are of particular importance in the organization of activities and are opposed to individual business functions or independent work areas. These teams can be used for identification, evaluation, and implementation of opportunities. Teams are empowered to take direct action, make decisions, and initiate changes that will result in continuous improvement of the energy system in order to comply with policies and achieve organizational goals (Ogbeibu, Emelifeonwu, Senadjki, Gaskin, & Kaivo-oja, 2020). Once the team has been established, it is necessary to clearly define the role of each individual in the team.

Identification of significant environmental impacts. The eco-management system is essentially derived from environmental problems and encourages organizations to systematically monitor the environmental impacts that result from carrying out their activities, products, and services. This approach can be very effective and can encourage organizations to take a proactive and sustainable approach to managing their environmental impacts and programs. Aspects of impacts are in fact elements of organizational activities, products, or services that may affect the environment. The organizational aspects may include generation of waste and pollution, use of resources, use and losses of energy, and other environmental impacts (Chen, Feng, & Ma, 2020). The standard proposes some basic planning activities that are used in most organizations in order to implement sustainable energy management, as follows:

- Identification of activities, operations, processes, products, and services that have environmental impacts;
- Identification of legal recommendations in the field of energy that influence organizational activities, products, and services;
- Evaluation of ecological impacts of energy consumption according to the degree of importance;

- Setting goals whose achievement reduces negative environmental impacts generated by the use of energy; and
- Selection and implementation of activities during the eco-management program that are necessary for achieving the desired goals.

During the procedures of identification and ranking of environmental impacts, planning can serve as a key element of the eco-management plan. Planned assessment is a systematic, periodic research of organizational operations designed to identify potential areas for change of energy performance. A properly designed eco-management system can have a far greater range than a traditional approach to energy management. In order to understand all environmental impacts, the organization should consider the use of energy and water, conservation of landscape, the problem of noise, and other impacts (Rabaia et al., 2021). In addition, an organization may request information from its information suppliers about the content of certain materials, the method and type of packaging, and methods of delivery. The procedure to identify aspects can be conducted by using the following tools (Barletta, Despeisse, Hoffenson, & Johansson, 2021):

- Process mapping
- Interviews
- Questionnaires
- Checklists
- Benchmarking
- Cost-benefit analysis
- Life cycle analysis
- Analysis of input-output balance

A particularly important aspect of determining environmental impact includes consideration of the state of the control systems in energy management. The organization should develop appropriate criteria for determining the degree of significance of these effects. Criteria may include regulatory activities and management costs and risks associated with the use of energy resources. The most important feature of these criteria lies in the fact that they reflect the organizational values stipulated in the business policy.

Identification of operations, as well as monitoring and measuring of activities that are associated with significant environmental impacts, leads to the creation of procedures for minimizing risks from environmental influences. A systematic approach encourages the organization to find new solutions, to identify different opportunities, and to surpass the traditional way of considering energy issues, according to which only people in charge of health and safety of employees deal with environmental problems.

Typically, organizations separate their own environmental impacts to the impacts on soil, air, and water, and compare them with current regulations. This approach leads to individual problems and application of the so-called *end-of-pipe* strategies that consider only the consequences and essentially lack efficiency and are linked with higher costs. Corrections should increase efficiency and reduce costs (Wu, Sun, & Xu, 2019).

The organization can have benefits if it includes its suppliers and partners into procedures of eco-management when it comes to important environmental impacts they can make on the environment. It can be necessary and useful to organize training for external colleagues if they directly influence environmental impacts. In this way, a forum of the two organizations is created to reconsider goods and services to improve them.

The organization may choose to modify the existing assessment tools or develop its own procedures to identify all organizational environmental impacts and the level of their significance. Using teamwork during the planning process, as well as sticking to the list of activities and time frame, is of particular importance. Possible improvements should be made first for the problems of the utmost importance. Finally, the importance of considering these issues beyond the organization and the existing regulations should be emphasized.

Identification of objectives and targets. The eco-management goals represent the most important part of the process of achieving the planned efforts to implement sustainable energy management (objectives and goals). *Objectives* are general environmental efforts that the organization wants to accomplish, while *goals* are detailed performances that are recommended in order to achieve the previously defined environmental efforts. Regardless of the specific nature of some goals, they have to be consistent with the existing organizational environmental policy.

Program techniques are most successful if achievable and measurable goals are set. In the stage of determining the goals, the organization should consider the significant environmental impacts, such as energy consumption, legal and other regulations, public attitudes, technical abilities, and flexibility of equipment. In this stage, the possibility of return on investment and business opportunities to correct its position in the market and gain profit should be considered. The eco-management approach encourages innovative solutions to the problems of using energy and materials at all levels of the organization. The objectives must be documented for all levels and business functions of the company that may affect the environment. The goals may be different at different levels such as the level of management, engineers, supervisors, and operatives (Duke, 2020). In short, the eco-management standard recommends that organizations set their own goals to reduce harmful environmental impact, choose the activities necessary to achieve their goals, and then carry out continuous monitoring in order to correct and improve the system.

Training and accountability. After defining the goals, it is necessary to conduct appropriate training. Employees must be familiar with their own environmental responsibility and therefore be trained to behave in an environmentally responsible manner. Depending on the situation, it is necessary to develop appropriate training that is required in order to achieve the desired objectives and integrate them into the existing environmental, health, and safety programs. Training should be provided to all employees in the implementation of sustainable energy management. Experience has shown that quality training of the participants is one of the most important factors for successful implementation.

Employees should be trained on the relevant procedures, their individual tasks and responsibilities, and be familiar with the expected results. After conducting training for all employees, the company will soon receive numerous proposals from employees about how to improve certain activities. Depending on the nature of the program, each company will predict the level of knowledge and training required for each individual to be able to achieve the environmental and energy requirements of his or her job (Lue, Byars, & Ibrahim, 2015).

Developing an energy program. The last element of planning within the eco-management system is the formation of an eco-management program. The program should contain an action plan, specified responsibilities at all organizational levels, the execution plan, timeline, and the resources required to meet objectives. Upon formation of the program, each individual who has been assigned certain responsibilities develops his or her own plan. Although the stage of achieving goals is separated from the stage of planning, they are closely related. It is necessary to have an idea in advance as to how the objectives can be achieved before the program is integrated into the company eco-management system.

An organization may have a number of eco-management programs. Sometimes a company develops eco-management programs for all determined objectives. In other cases, a company establishes specific eco-management programs for regulatory aspects, solid waste, energy use, etc. Finally, each company should follow the implementation of eco-management programs and make adjustments as needed.

Implementation

There is a clear need to conduct a careful evaluation of the ways in which the implementation of an eco-management system will affect organizational energy performance. At this point, it is necessary to decide how to incorporate the eco-management approach into the existing environmental regulations. The stage of implementation of eco-management system can be represented as shown in Fig. 3.12.

The first substage of implementation involves the development and control of documents. Today, most organizations have already developed procedures for monitoring energy processes, which include work instructions, training plans, the results of testing and monitoring, and calibration instructions. It is important that organizations create their own documentation whenever appropriate (Gopinath, Kumar, Joshua, & Srinivas, 2020). The documentation provides an answer to the question of how to perform certain actions and provides step-by-step instructions for task execution. System requirements necessary for the operation have to be documented. The documentation should be as clear and as simple as possible.

If the instructions and the documentation do not add some value to operational control, are they necessary at all? Detailed documentation is not required for all departments in a company. The main factor affecting the need for documentation and its scope is the risk and complexity of activities as well as the frequency of control necessary to perform a particular activity. The organization team needs to identify differences in relation to existing documentation and, if necessary, initiate the creation of new procedures.

Operational control develops procedures for performing certain activities and defines the operational criteria as well as preventive and corrective measures. Implementation of operational control is in the true sense the working part of the eco-management system. Procedures are instructions that organizations use to implement environmental activities. Procedures determine who, what, when, where, and why for some of the planned activities. At this point, most organizations face a great number of problems. Since the developed procedures are extremely important, the organization

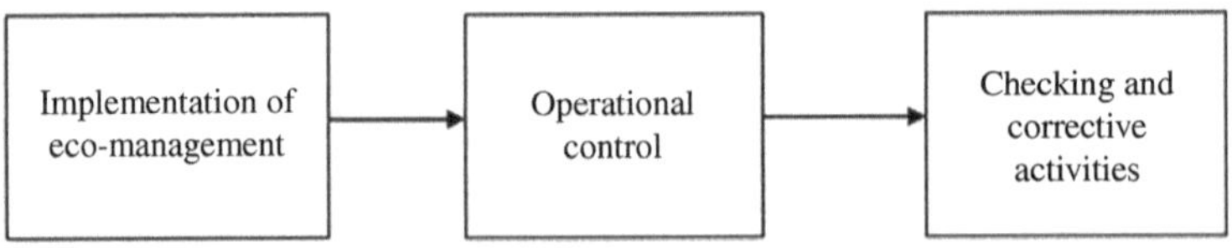

Fig. 3.12 Implementation of eco-management system.

will benefit if it properly determines what procedures will be documented and, if it documents them, the form of guidelines for training and implementation. Written procedures are an essential element of operational control, and if they are lacking, this often leads to large deviations and discrepancies (Antony & Govindarajan, 2017).

An important step in the phase of operational control is the identification of individual business activities and functions that may have active or potential impact on the environment. Operational control that is specified for the most important environmental impacts helps the company define the roles, responsibilities, and authority necessary in order to achieve the desired performance. Quality operational control can be briefly defined as a procedure or a process within an organization that reduces the generation of waste and preserves natural resources.

If, for example, there is a change in the working process in which the waste is generated—for example, air pollution due to using fossil fuels—it is necessary to adequately modify operational control. For example, by appropriately marking materials and waste, it is possible to reduce the risk of incorrect disposal of substances. If it properly separates and disposes the particular elements of waste, the organization can access the recycling or reuse of materials (Kochhar, 2012). Operational control should also ensure the proper use of equipment. In order to avoid possible failures and inefficiencies, preventive measures should be developed and implemented.

After the substage of operational control, there is the substage of checking and implementation of corrective actions. It is necessary to control each active system in order to see whether the system is working well, to identify current or potential problems, and to determine ways in which identified problems can be solved. Certain measures of environmental performance have been specified and, according to the results, corrective actions are taken if needed.

Evaluation and observation

The eco-management system requires certain procedures in terms of monitoring and measuring energy performance, collecting information needed to monitor and assess the potential for achieving the goals, and the assessment of compliance with environmental regulations. A particularly critical moment is the determination of what should be monitored and measured and which information needs to be collected. As previously set goals are measurable, the whole process leading to their implementation must be

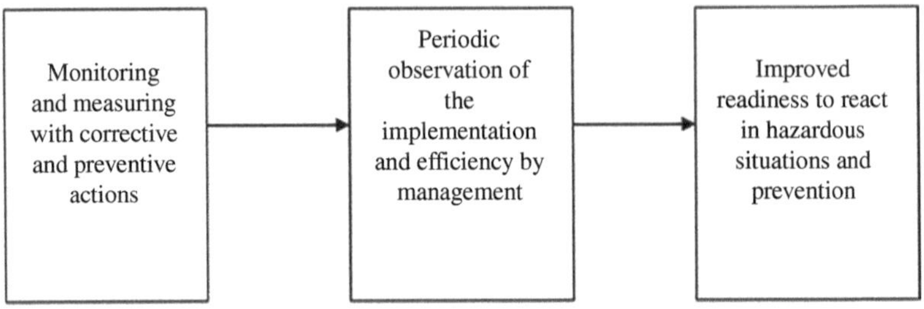

Fig. 3.13 Evaluation and observation of the eco-management system of sustainable management of energy.

monitored. The evaluation phase and observation of the functioning of the eco-management system are shown in Fig. 3.13.

Eco-management measurements are used as indicators of energy performance and must be in accordance with legal and other regulations. The number of indicators that will be established and monitored must be carefully determined. A large number of indicators may produce a huge amount of information that surpasses the abilities of the information processing system, while an insufficient number of indicators do not provide enough information needed for effective decision-making. In order to conduct high-quality measuring, it is necessary to determine who is responsible for monitoring, analyzing, collecting, and processing data, the frequency of measurement required to obtain quality information, the way the data will be analyzed, and the way in which the data will be presented (Beisheim, Krämer, & Engell, 2020).

A company may do monitoring every day or every hour, depending on needs. Monitoring will be used to determine the existence and nature of the trend so that deviation can be quickly and easily identified. Monitoring of eco-management performance is specific in that it requires monitoring at different levels of work and business functions, i.e., it is focused on employees and their effects.

The review and improvement by management are conducted in order to provide continuous improvement, which is one of the basic characteristics of this approach. Top management should periodically examine the implementation of the eco-management system and its efficiency. Experience has shown that the efficiency of management directly affects the chances of successful implementation of the eco-management system. The eco-management system is a business system that directs the organization to manage its environmental issues in a way that is based on continuous improvement. The eco-management system is focused primarily on the support of top management and involvement of all employees.

If discrepancies are identified during observation, the person responsible should determine the way in which adjustments will be implemented and how the system of prevention will be developed so as not to allow similar situations to repeat in the future. Readiness to respond in emergency situations is a necessary step because these situations can happen even in the best-prepared organizations. In order to reduce the probability of such events to a minimum, the eco-management system develops a system of procedures for the implementation of preventive measures.

Studies on the application of preventive and corrective measures and the ways of reacting in emergency situations must be integrated into operational control procedures. Procedures may include various changes, from the way of thinking of managers to changes in direct production. Preventive procedures should be perceived and updated when needed after an environmental accident or emergency situation.

Total quality management approach

The realization of the process of sustainable energy management is often considered a job to be dealt with by the staff responsible for ecology. It is considered much more effective if the program is integrated into overall organizational business practices. This approach allows the engagement of both top management and employees at all levels. In this section, the most commonly used model of quality management will be described for the purpose of implementation of the sustainable management of energy, which is based on the *Baldrige Quality Program*.

The *Malcolm Baldrige National Award* was established by the US President and Congress in 1987 in an effort to improve the quality of business in American companies. However, the criteria for the award soon surpassed American borders and became an example for quality improvement in the global market. Currently, the *Baldrige Award* is given in almost 50 countries around the world.

The first well-documented example of use of quality management was created by the integration of the quality model in order to monitor the environmental performance, dates back in 1998, and was applied as the *Green Zia* program in New Mexico. It was then that the criterion of environmental excellence was established for the first time (with the analogy to the existing criterion of performance excellence, defined by the quality model). Environmental excellence can be defined as the ultimate goal that can be achieved by applying the quality model in the management of EHS (Environment, Health, and Safety) management.

An environmental excellence program establishes an appropriate range of objectives that can be achieved by reaching the best status in those areas that best support prevention-oriented approaches to EHS management. The results that can be achieved by implementation of goal-oriented activities are the output of EHS performance but not the measuring of the performance itself, and thus organizations cannot indicate whether their own results are better than the results achieved by their competition. By focusing on performance, an organization can contribute to the development of the program and provide a means of measuring their efficiency in terms of the functioning of the process of sustainable energy management.

Using the quality model, the range of goals can be defined as *zero-zero defects* (zero rejects), *zero equipment breakdowns* (zero equipment failures), and *zero waste* (zero waste). Many organizations today have expanded the zero concept of their EHS program to zero incidents, zero accidents, zero emissions, and zero drain—extinction of the world's resources, which is the main precognition of sustainability).

The *Green Zia* model has all the elements necessary to realize the vision of zero waste. First, the criteria were given to define what is best in the class, allowing each organization to be able to measure its own progress toward achieving excellence. Second, the model offers a guide that can be used for integration of the criteria into the program. Third, it includes a rigorous summative system that measures the results of environmental excellence on the scale of 1000 units. Fourth, it provides a feedback report on the strengths and weaknesses when applying the program for the achievement of environmental excellence in comparison to the criteria given in the basic principles (guidelines). The summary report and the feedback report represent important tools for the organization trying to move toward achieving environmental excellence. Organizations achieve excellence in their own way by applying this model. They choose the performance they want to improve and determine how to incorporate the achieved results in the total efforts of the company.

In practice, the two models are commonly used—one contains seven criteria and the other contains eleven criteria.

The seven criteria of the quality model

Since the implementation of the program of sustainable energy management is a complex process, the most important factors must never be ignored. Based on many years of experience in applying quality management, the seven basic criteria have been defined, as follows:

- Leadership;
- Strategic planning;
- Involvement of stakeholders;
- Involvement of employees;
- Process management;
- Information analysis; and
- Results

These criteria are incorporated into the basic principles of the *Green Zia* program and are used for the purpose of application of the quality model in the implementation of the energy management program. Based on these criteria, the organization will be able to determine how to integrate the energy management program into its own business. In the past, organizations mainly received information about why something needed to be done, but not how to achieve it.

Leadership. Intensive use of *top-down* direction in the application of the quality model increases the chances of success of the program and helps to integrate it into the organization as a whole. A sustainable energy management program will be considered to the extent that the leadership of the organization considers it necessary and important. At this point, it is necessary to raise two basic questions:

- How can management of the company transfer its commitment to its employees in terms of implementation of sustainable energy management?
- How can management demonstrate its determination and commitment?

The answers to these questions can only be obtained through discussions with managers, which can often take too much time.

Strategic planning. Management of a company usually uses a specific form of strategic planning in order to lead the organization. Sometimes it uses a certain strategic plan as well. In other cases, strategic planning can have a less formal form. The basic questions that arise during strategic planning include:

- How do we identify long-term and short-term aims and intentions?
- How do we develop these aims and intentions?

- How do we apply achieved aims and intentions?
- How do certain aims and intentions influence general business goals and intentions of the organization as a whole?

Ideally, the sustainable energy management program will be considered important in the minds of managers and an important part of strategic planning. There is a strong link between strategic planning and leadership.

Involvement of stakeholders. Organizations do not operate alone. There are many other organizations that have an impact on the implementation of sustainable energy management. Interested parties include a wide range of stakeholders such as consumers, suppliers, contractors, regulatory bodies, NGOs, environmental associations, social associations, and the general public. This involvement can be understood by raising the following questions:

- How does the organization involve the stakeholders in the development and implementation of the program of sustainable energy management?
- Is the organization involved in the program of sustainable management of energy in other organizations, and if so, how?

The greatest attention should be paid to the analysis of the influence of the employees as stakeholders.

Involvement of employees. This criterion analyzes the bottom-up orientation of the program of sustainable energy management, which is as important as the top-down orientation, included in the criterion that analyzes the attitudes of management. Employees are a very important element of the program because they know best the details of the business operations of the organization. The following questions should be answered:

- How can the organization prepare and involve employees in the development and implementation of the program of sustainable management of energy?
- In what ways are the values and well-being of employees involved in the program of sustainable management of energy?

Process management. This is the criterion by which the ISO 14001 standard or other eco-management standards help the organization implement sustainable energy management. Process management considers how to manage work processes in a manner established by the program. Issues to be considered include:

- How does the organization identify the basic and auxiliary work processes that influence the program of sustainable energy management?
- How does the organization analyze these work processes in order to understand their causes and effects?

- How does the organization manage work processes to achieve program excellence?

The criteria stipulated above are closely linked with the criteria of information analysis.

Information analysis. The *Green Zia* program is, overall, based on the fact that all criteria are mutually linked. Each time one criterion is taken into account in order to make a change, it is necessary to consider how this will affect the other criteria. It is important to emphasize that it is not necessary to be perfect in all seven criteria of the Quality Model, but to understand that some criteria will perform better or worse than others.

Results. The above seven criteria of Quality Model will enable the creation of images of the situation in the organization and, grounded on that, certain results of the analysis will be obtained. They will reveal the weaknesses, strengths, and improvement opportunities.

The eleven criteria of the quality model

Principles that are often referred to as core values are used in the context of all activities of an organization. They are designed to provide a guide for decision-making at all levels in an organization. Each organization should choose the way in which each of these principles is going to be integrated into relevant criteria in the quality model to implement the program of sustainable energy management in its business. If the organization already has the basic principles in place, it is necessary to consider the application of sustainable energy management program in all these areas.

If the organization does not have the previously defined basic principles, it should first become familiar with the way in which they can first become a part of the organizational culture. In this case, the initial activities may take longer because it is necessary to implement the changes that have to become an integral part of the organization's culture. Experience has shown that it takes at least 2 years to engage in certain activities, including incorporating environmentally responsible activities in the core business values. When this is achieved, the integration of sustainable energy management can begin. Within the quality model, it is necessary to consider the following 11 criteria:

- Stakeholders-oriented program;
- Leadership;
- Continuous learning and improvement;
- Evaluation of employees;
- Fast response;

- Effective design of products, services, and processes;
- Long-term planning;
- Facts-based management;
- Partnership development;
- Public responsibility; and
- Focus on results.

Stakeholders-oriented program. A sustainable energy management program evaluates all interested parties, so it must be expressed through all products or services whose characteristics meet the values of stakeholders.

A sustainable energy management program oriented toward the stakeholders is thus a strategic concept. It focuses on customer retention, conquering a new part of the market, growth, and ensuring long-term communication with the environment. It requires a constant willingness to change the requirements of stakeholders and the market as a whole in terms of facts affecting their satisfaction.

A program oriented toward all interested parties is much more than a simple reduction in the amount of waste and reduction of emissions into the environment, meeting the statutory provisions, or reducing the number of complaints. On the other hand, minimization of waste or elimination of other causes of dissatisfaction of interested parties is a particularly important part of the program. The success of an organization in terms of recovery from possible EHS issues is key in establishing better and more stable position of organization in its environment.

Leadership. Top management of an organization is a team of people who create and direct the focus of interested parties and create clear and visible values and high expectations. The above directions, values, and expectations should be directed to all stakeholders. Leaders can ensure the creation of strategies, systems, and methods for achieving environmental and energy excellence. They should stimulate innovation and promote knowledge and skills. The strategy and values will help manage the activities and decision-making in the organization. Managers who are focused on the development will encourage the participation of all staff by teaching all participants to be innovative and creative.

Based on the behavior and the role in the process of planning and communication and review of performance, leadership serves as a kind of model of behavior and actions, which gives a stimulus to activities in the whole organization.

Continuous learning and improvement. Achieving the highest levels of energy performance requires high-quality design of the executive approach to continuous improvement and learning. The term continuous improvement refers to the improvement of existing activities and the adoption of completely new solutions and major changes. The term implies willingness to change, moving toward new goals and approaches. Improvement and learning need to be a regular part of daily activities. They should be practiced at individual and all other organizational levels and should be oriented toward creating opportunities for innovation and better functioning of sustainable energy management. Improvement and learning include increasing the value expected by stakeholders through new and improved products and services; developing new business opportunities; reducing waste, emissions, and costs; improving the level of responsibility as an activity without creating a direct revenue; increasing efficiency in productivity by using all resources (energy, water, and raw materials); and increasing the value of organizational performance in order to comply with the full social and civilizational responsibility. Therefore, improvement and learning are directed not only to better products and services, but also to increase the level of environmental responsibility.

Evaluation of employees. Success in the implementation of sustainable energy management in an organization depends largely on the knowledge, skills, innovation, creativity, and motivation of employees. The success in the work of employees largely depends on the ability to get a chance to learn and apply new knowledge and skills. An organization can gain considerable advantage by using proper work potential through investments in learning, training, and creating opportunities for continuous improvement. Adequate training provides a very effective way to improve work processes and training on new technologies. Education and training should be adapted to the requirements of each job. The main challenges of evaluating the performance of employees are integration of human resource management activities (selection, performance, recognition, climbing the career ladder); development, education, and transfer of knowledge of employees; and harmonization of the human resources management with strategic processes. The answer to these challenges requires monitoring of the knowledge, skills, satisfaction, and motivation of employees in terms of sustainable energy management. Data on the above can serve as a useful indicator of the conditions in the organization or in some of its parts.

Fast response. Obtaining permissions and compliance with the regulations may require additional time in the decision-making process. Success in the global market requires shorter cycles for introduction of any improvements and innovations. Also, faster and more flexible response to requests from stakeholders becomes increasingly important. Great progress is made in terms of time needed for responding to the requirements toward simplification of work units and processes together with the incorporation of the program in the design stage (the so-called *design for environment*). Other time improvements are related to the fact that the time required for improvement leads to simultaneous changes in organizational behavior, quality, productivity, and costs.

Effective design of products, services, and processes. Organizations need to integrate the program of sustainable energy management into the design stage itself. Experience has proven that the program that is integrated at this stage is much more effective than if integrated into the processes that occur in the stage of production or even that stage. The design includes the creation of such products, services, and processes that will be environmental and energy-friendly by their nature. Costs arising from preventive problem-solving are much lower than the costs that arise after the problem has been caused. The design stage is critical and demanding, especially in terms of social responsibility.

Long-term planning for the future. Survival in the market implies a strong orientation toward the future and the willingness to take on long-term obligations toward all stakeholders. Organizations should assume and predict a number of factors in their own strategic plans, such as expectations of stakeholders, new business opportunities, changes in the global marketplace, technological developments, new market segments and new consumers, changes in legal regulations, social expectations, and strategic efforts of competition. Short-term and long-term plans of sustainable energy management, strategic objectives, and allocation of resources should reflect these influences. The most important factor that takes into account the long-term nature of obligation is the development of the awareness of employees and suppliers that they are the key participants in the process.

Facts-based management. Organization is largely dependent on the measurement and analysis of energy performance. Each measurement is determined by an organizational strategy and provides important information about key processes, outputs, and outcomes. To measure energy performance, different types of data and information are needed. The areas of observation include satisfaction of all stakeholders, operations, market position, and financial indicators. Only on the basis of the collected facts should management make its decisions.

The analysis involves extraction of the basic meaning of the energy data and information that can help evaluation, decision-making, and improvements in an organization on the basis of established trends and projections, which are impossible to reach without adequate analysis. Identified trends and projections are of particular importance for understanding the progress of the process and can serve as a starting point for benchmarking the results of the competitors.

Partnership development. Experience has shown that organizations can quickly and efficiently achieve their desired goals by building appropriate internal and external networks and partnerships. Partnerships are more desirable in the modern world, where energy responsibility is one of the pillars of corporate social responsibility.

Internal partnerships might include cooperation of employees in the same area, including mutual training, sharing experience, and creating teams. In addition, internal partnerships encourage the creation of internal links between business units, improving flexibility and acquiring new knowledge.

External partnerships can be built with customers, suppliers, NGOs, environmental associations, and educational institutions for various reasons, among which the main one is training. A particularly important form of external partnership is a strategic partnership (strategic alliances). A high-quality external partnership can help with entering new markets. Strategic partnerships allow combining of organizational core competencies and leadership skills with complementary strengths and capabilities of partners, thereby resulting in the increase in total capacity of all partners. Internal and external partners can develop their own long-term energy-related goals and, based on them, plan and implement joint investments.

Public responsibility. Today, organizations gain a special advantage if they prove their environmental responsibility according to the demands of society and behave as civilized citizens. Responsibility is primarily related to an organization's core values, which should include health care, safety, and the environment. In this way, organizations create a positive image and gain the loyalty of consumers and society as a whole. Particularly important aspects that should be considered at this point are the promotion of education, promotion of health care in the community, improvement of environmental quality in the region, and participation in local activities that improve quality of life.

Focus on results. By measuring energy performance, it is possible to get the proper results and meet the requirements of stakeholders. Results that do not meet the stipulated requirements suggest the need to make corrections without delay. With this approach, the organization is constantly in need of balancing between its own abilities and requirements of interested parties, without jeopardizing the requirements of any interested party. This is possible only if the organization focuses on the energy results, which will be the only measure of success of implementation of sustainable energy management.

Application of the quality model in realization of the pollution prevention program

For the purpose of the integration of sustainable energy management into organizational practices, the application of the quality model, which contains the five basic steps, is given in Fig. 3.14.

Planning and development of the strategy of sustainable management of energy

This step can be summarized as shown in Fig. 3.15.

Planning the strategy of sustainable energy management. The first step in application of the quality model involves determination of the elements of the

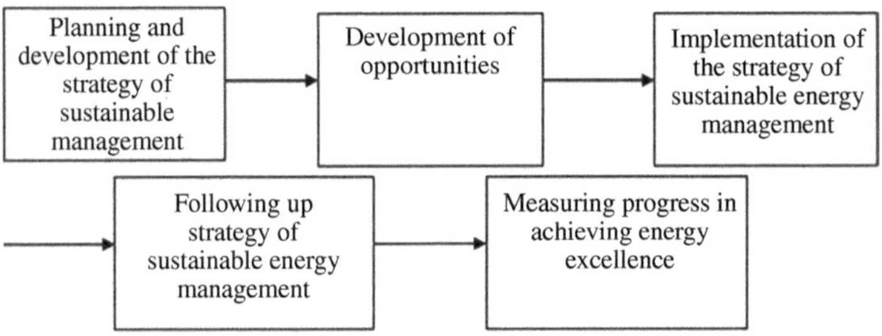

Fig. 3.14 Basic steps in application of the quality model in the implementation of the strategy of sustainable management.

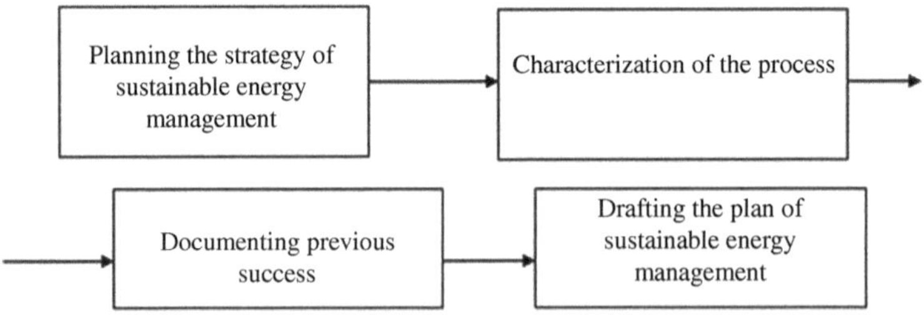

Fig. 3.15 Planning and development of the strategy of sustainable energy management.

quality model, which refer to planning of the strategy of sustainable energy management. This primarily refers to strategic planning, focusing on the demands of stakeholders, leadership, and involvement of employees. Furthermore, it is necessary to identify inconsistencies and deviations (gap analysis) that may arise between the detailed issues that are anticipated by *Green Zia* program for each of these criteria and what is currently happening in the organization. Gap analysis should also consider how the strategy of sustainable energy management can be integrated into the business practice by using certain combinations of the presented 11 principles of the quality model.

Furthermore, it is necessary to consider the element of leadership, which helps in preparation of the policy of sustainable energy management. Examination of leadership will point out how top managers demonstrate their own commitment to the need for continuous environmental improvement to employees and other stakeholders.

Including employees helps in preparation of the training program for employees because its consideration brings insight into the condition of interest and training of employees for the implementation of the program of environmental improvement. In addition, an insight into their expectations will be made in terms of improving health and safety.

Strategic planning is a criterion that indicates how the organization identifies, develops, and implements long- and short-term goals and efforts in the field of continuous improvement of the environment and how the goals are related to general business aims, mission, and vision of the organization.

Finally, the criterion of stakeholders determines whether and how the organization involves stakeholders in the development and implementation of the program of environmental improvement.

Characterization of the process. At this point, it is necessary to characterize the energy processes that take place in the organization, and for this purpose, the most useful is the application of the process mapping that is envisioned in the system approach to implementation of pollution prevention. The determined process maps will be used as a template for information in the process, which will consequently identify production units that require further analysis.

Besides the mapping process, it is necessary to review the criteria of the process management and information analysis. The analysis of the process management criteria will provide information on how the organization identifies, analyzes, and manages processes that potentially affect the environment or may cause injury and illness of employees. Analysis of information provides insight into how information that affects the efficiency of the realization of the program and decision-making process is selected, collected, and used in the organization.

It should be noted that six out of seven criteria and all 11 basic principles of the quality model are contained in the first two steps.

Documenting previous success. It is best to document the previous activities in the field of pollution prevention. Regardless of the fact that most organizations will be at the beginning of implementation of sustainable energy management, it is necessary to note what has been done so far in the preservation and improvement of the quality of the living environment. The time frame that should be examined entails the previous period of 2–5 years. Employees and management should be familiar with the current activities and try to build in all positive experiences and steps into their future program of prevention of pollution.

Drafting the plan of the sustainable energy management. Having implemented activities, it is necessary to prepare a draft plan in terms of pollution prevention, which will later be discussed by the internal and external stakeholders.

Development of organizational possibilities

The planning stage of sustainable energy management is followed by the determination of the opportunities for improvement within the program, which can be summarized as shown in Fig. 3.16.

Collection and analysis of information. A hierarchical process map, prepared in the previous stage, is used to collect information in defined production units. After that, the team will be able to identify and develop opportunities. If the analysis, for example, identifies the use of toxic materials, it is an opportunity for improvement to avoid the use of toxic materials. The analysis can provide a huge number of potential opportunities for improvement.

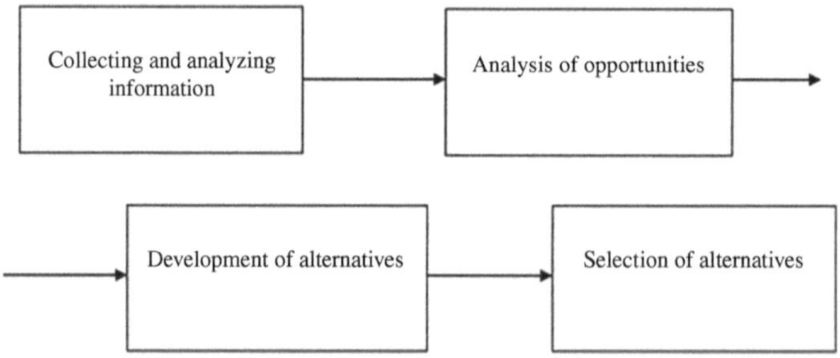

Fig. 3.16 Development of opportunities of sustainable management of energy.

It is necessary to rank them by application of so-called Pareto analysis. It is optimal to get 8 out of 11 opportunities for improvement at this stage. These options should be selected in accordance with the objectives to be achieved in 1 year. The program team should develop their own criteria for a fast analysis. If necessary, the team can undertake further analysis of one or two opportunities, which may take more time but should not last longer than 2 years.

Analysis of opportunities. This step begins with an analysis of the production units. For any preselected opportunity for improvement, the team discusses possibilities of implementation, which is most often done through the resource analysis, methods, equipment, and technology, as well as people who will be involved in the implementation of the proposed options. After the analysis, the team can provide a final evaluation in terms of resolution of the problem and suggest a certain number of alternatives.

Development of alternatives. Consideration of alternatives is the process of selection of the proposed solutions in order to choose the solution with the best possible outcomes.

Selection of alternatives. After consideration of the alternatives, it is necessary to select them based on the analysis defined in a system approach to the implementation of sustainable energy management.

Implementation of the pollution prevention program

The first step in the implementation of sustainable energy management is the development of an action plan for all the alternatives recommended in the previous stage. The action plan is an essential part of a sustainable energy management strategy for each year and must be carefully considered prior to application. It is best to reconsider the relevant steps of quality models that are evaluated during the first phase of the program, which is crucial to the success of pollution prevention.

The action plan should be revised at least quarterly. After 1 year, a complete internal and external evaluation should be done. Only after a complete evaluation is the pollution prevention program ready for implementation.

Following up on the strategy of sustainable energy management

Throughout the implementation phase of sustainable energy management, it is necessary to understand it thoroughly in order to obtain data needed for the analysis and summary of results. The criterion result (the seventh criterion of the quality model) examines the actual situation in the organization and provides the results of the strategy for sustainable energy management. In addition, it is recommended to consider the levels and trends and the ways in which they refer to the impact on the environment, health, and security of employees, the requirements of other interested parties, and the main financial indicators.

It is important to stick to the rule that only what has been measured has been achieved. A pollution prevention program cannot be based only on the expression of wishes. The achieved results should serve as progress indicators and not only as the data that will be presented to the stakeholders. After reporting on the results, the action plan can be developed for the next year.

Measuring progress in achieving environmental energy excellence

After several years, the strategy of sustainable energy management can be considered in terms of the contribution provided, i.e., the achievement of organizational goals. The results can be summarized by the proposal envisioned by the *Green Zia* program. In this way, the insight will be included in the pollution prevention program for the entire organization. The organization will also be able to compare the achievements of its competitors, and corrections can be made as needed.

Life cycle analysis

Life cycle analysis (LCA) is currently the most common method of designing environmentally friendly products and technologies and evaluating their impact on the environment. This method can also be successfully used for the implementation of certain aspects of the strategy of sustainable energy management.

This method is governed by the international series of ISO 14000 standards, which is also the reason why it is widely used in relation to other methods of designing for the environment. LCA helps companies realize the extent of the impact of their products, processes, and other activities on the environment. The goal of LCA is not only to provide an answer to the question of how serious the harmful effect is, but also to enable strategic planning of future activities.

LCA history dates back to the early 1960s, when scientists became concerned with the exhaustion of fossil fuels. LCA methodology was initially developed in order to get insight into the effect of production, distribution, and consumption of the fuel on the environment. From 1970 to 1975, due to the oil crisis, several analyses of profile resources and the impact of their exploitation on the environment (called Resource and Environmental Profile Analysis) were conducted.

During this study, the protocol was developed as well as the standard methodology for conducting such studies. After considering the impact of energy production and consumption in the late 1970s, LCA focused on the analysis of the life cycle of waste.

Basic steps of life cycle analysis

LCA is by definition a systematic approach to evaluating the environmental consequences of a particular product, process, or activity from the acquisition of raw materials to final disposal. The basic steps of LCA are given in Fig. 3.17.

Raw materials acquisition. Each product or service is made by using certain types and amounts of energy. Therefore, each production process takes from nature certain materials, and the exploitation of energy is one of the most intensive forms of exploitation of raw materials in general. Regardless of the type of activity that a company deals with, it uses certain amounts of energy. A great deal of energy need is met with conventional energy resources, whose exploitation and consumption are usually connected to a lot of pressure on environmental quality, destruction of huge areas of land, and other forms of pollution. Every company has to take into account the damage that is caused to the natural environment because of its energy needs.

Production. In the broadest sense, every production process involves a certain amount of energy consumption, and the amount of energy used greatly depends on the type of activity, as well as the quality of technology that is used, the intensity of its use, and the skills of the producers. Estimation of energy consumption in a production system gives managers insight into the (non-)compliance of sustainable development and is one of the key areas for the implementation of improvements.

Distribution. Most goods in contemporary business are produced at some distance from the place of consumption and use. Therefore, the distribution

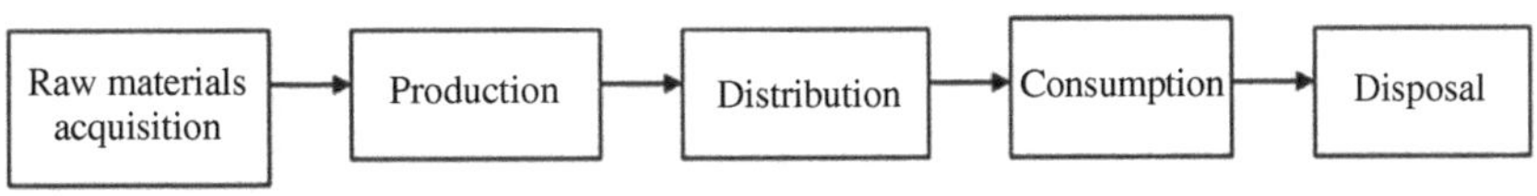

Fig. 3.17 Basic steps of the life cycle analysis method.

is associated with huge amounts of energy for transport. Speed and distance from the destination, the way of transport, and the use of alternative energy sources for these purposes are not taken into account. Therefore, the estimation of the specific energy consumption for the needs of distribution is of great importance, because opportunities for saving energy are limited.

Consumption. Each product is spent for a particular purpose in a certain way and thereby requires more or less energy consumption. The estimation of the energy efficiency of certain products is particularly important in the category of products, one of the larger consumers of energy. The intention to produce goods that use less energy is the imperative of contemporary business and market requirements.

Disposal. After being used, the product that is not consumed as a whole, or after the consumer decides to stop using it, needs to be disposed, having lost the purpose for which it was obtained. The possibility of safe disposal is certainly of special interest and is subject to numerous efforts to create products that can be disposed of in a manner that does not harm the environment and does not require intensive consumption of energy in that process.

LCA is a broad term that can be applied to a number of activities, but on the other hand, its comprehensiveness is limited. Namely, it usually does not provide data of satisfactory quality and the quantity of data is often questionable. Additionally, accurate measurements are not provided, and attention is not paid to specific forms of energy changes during work processes. Therefore, LCA is well accepted exclusively as a method for the assessment of the energy load of a product. The application of this method for description and assessment of the production cycle is extremely impractical and inefficient.

The Main components of life cycle analysis

LCA is used to assess the impact of each part of the process of obtaining a certain product. The basic components of LCA are:

- Inventory analysis;
- Estimation of impact; and
- Improvement possibility analysis.

Inventory analysis

Analysis of inventory is considered the most objective component of LCA, because it is based on data obtained by measuring the amount of energy and raw materials, air emissions, liquid effluents, solid waste, and all other processes in the life cycle of products, technology, or activities. By doing inventory analysis, it is possible to:

- Establish a set of all system requirements for energy resources;
- Identify the components of the process on which energy resources consumption could be reduced;
- Assist in developing new products or processes in which needs have been reduced in terms of resources and the amount of emission into the environment;
- Compare alternative materials, products, or activities; and
- Compare internal inventory information at other producers or products in the same or other company.

Inventory analysis within LCA can be schematically presented as shown in Fig. 3.18.

Given the above scheme, which is very simplified, it is necessary to know that the system of inventory analysis includes the collection of a large number of diverse data, where one should always keep in mind that the data will not be fully and accurately collected because it is influenced by a very large number of factors.

Acquisition (extraction) of raw materials is the first step in any production process, and therefore it is the first to be analyzed. Input in each production system represents a large number of components, so each input component that accounts for less than 5% can be neglected in the analysis. Otherwise, the analysis of inventory will be even more complex, more expensive, and harder to use. Properly conducted analysis of inventory will provide the ratio of the environmental impact of raw material acquisition. When analyzing

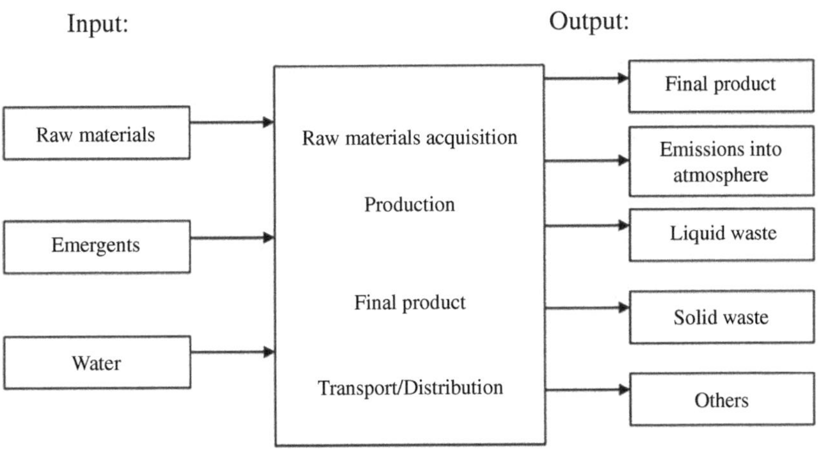

Fig. 3.18 Inventory analysis.

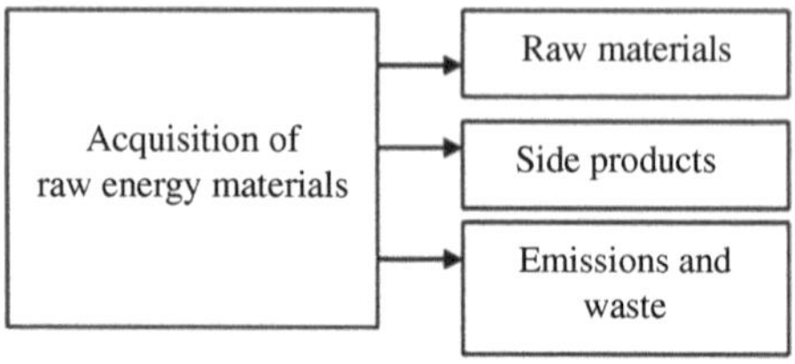

Fig. 3.19 Acquisition of raw materials.

the acquisition of raw materials inventory, it is necessary to specifically monitor the input materials and energy, as shown in Fig. 3.19.

Production, as the next stage, consists of a series of special processes (subsystems) so that each process can be separately analyzed. Often in the production process there is one major and several minor systems. In the analysis of inventory at this stage, the data is obtained on the movement of the substance and energy through the production process, as well as data on emissions into the environment. In addition to the analysis of the manufacturing process, this part of LCA involves analyzing the environmental impact of packaging as well as storage and distribution (Fig. 3.20).

Transport and distribution are analyzed separately for a number of environmental impacts. Environmental impacts are caused by fuel consumption during loading, repacking, transport, and distribution to the point of delivery. In this analysis, the transport distance is measured as well as the contents of cargo and the protection of the environment during transport. The transport and distribution system to be analyzed can be represented as shown in Fig. 3.21.

Consumption and disposal of products are observed and analyzed together and refer to a part of the system that arises from the moment when the product reaches the stage of consumption until it has been disposed after use. The inventory analysis phase aims to review all energy and other aspects of this stage of LCA, as the products are consumed in different ways and thus behave differently. In addition, there are different reactions during the

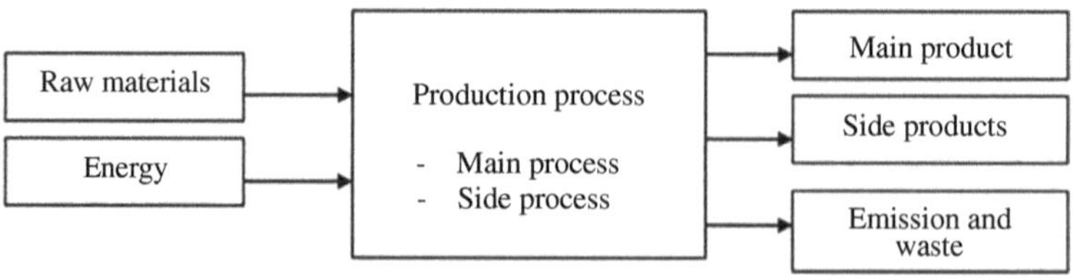

Fig. 3.20 Process of production analysis.

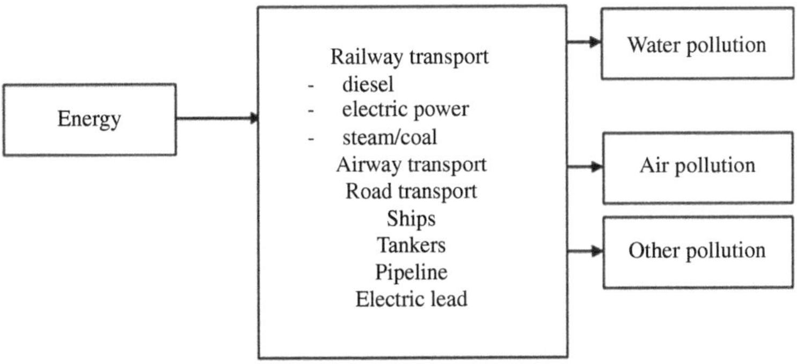

Fig. 3.21 Analysis of transport and distribution.

proper or improper use of the product. The main activities of consumers covered by this analysis are the use of the product and its storage, preparation, handling, maintenance (repair), and reuse. The main issues discussed at this stage include:

- The time the product was used prior to its rejection;
- The effects of the use of the product;
- The input parameters necessary in the maintenance process;
- The frequency of repairs; and
- The potential possibilities of reusing the product.

The system of consumption and disposal of products after use can be summarized as shown in Fig. 3.22.

During the implementation of these activities, there is energy consumption and therefore an impact on the environment. Only products that can be completely recycled in a closed system of recycling (in which no amount of waste is created) can be said to have zero effect on the environment.

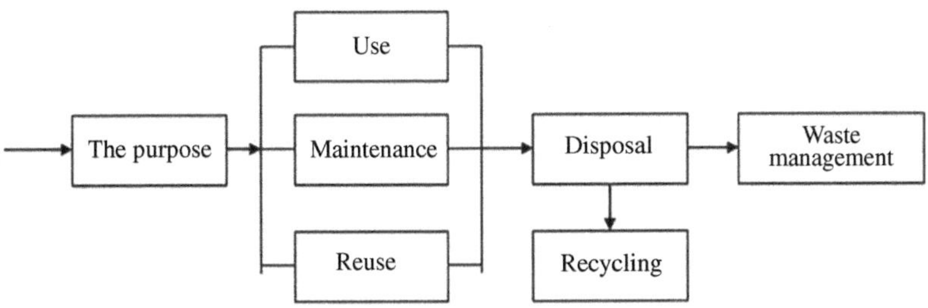

Fig. 3.22 Analysis of consumption and after use disposal.

Estimation of impact

After collecting data during analysis of the inventory, the estimation of its impact on the environment is conducted. The environment is complex, and the impacts may be numerous and varied, so this stage of LCA is particularly complex. However, this stage of analysis tells how certain stages in the product life cycle have harmful effects on the environment and gives concrete suggestions as to what can be done to improve the situation.

It should be noted that this phase of LCA is associated with a number of objective difficulties in implementation, because the estimation of the environmental impact itself is very complex and often there are no valid and reliable data on specific impacts. Nevertheless, the evaluation of the impact of the effects on the environment should be carried out at least for those parameters whose effect has already been studied and well known.

Analysis of improvement possibilities

Based on data collected in the inventory analysis and evaluation of their impact on the environment, at this stage suggestions are given to improve the life cycle of a possible product or process, or to improve the environmental identity of companies. There are many different ways to improve the energy image of a company, which later results in a number of positive financial and nonfinancial effects. Suggestions for improvements are determined depending on priorities, possibilities of implementation, and environmental and economic cost benefits.

Main problems in implementation

As discussed, LCA is the oldest and most accepted method of designing for the environment and is regulated by the international system of standards of environmental protection. Nevertheless, the application of this method has certain limitations, and there are many difficulties in its implementation. The most common problems in the application of LCA include:

- Increased costs of application in small enterprises;
- Time needed to master this method can surpass time of the life cycle of the product;
- Sometimes it is difficult to define time and space dimensions of the observed system;
- Often it is difficult to compare alternative solutions due to different parameters being compared;
- Sometimes it is necessary to analyze a large number of parameters;
- Data collection can be difficult and inaccurate; and

- It is difficult to objectively measure influence of the effects on the ecosystem of some harmful emissions, and it is even more difficult to be objective when measuring human health, now and in the future.

All these difficulties and limitations should be kept in mind when deciding on the implementation of LCA, but its advantages should never be ignored, and it should be applied whenever it is possible and economically justified.

Gap analysis

For the purpose of sustainable energy management, gap analysis is suggested as a very efficient technique. Gap analysis is one of the techniques of strategic management and forecasting used to determine the opportunities for growth and development in specific areas. This method is based on a number of estimates and assumptions, and its simplified schematic diagram is shown in Fig. 3.23.

In traditional terms, gap analysis is used to project long-term goals and determine the gap between projected goals and objectives that the company will probably meet if no new management activities are undertaken in relation to the current ones. Successful implementation of gap analysis involves a complex and comprehensive process of analysis and assumptions. For the purpose of sustainable energy management, these requirements are further defined in accordance with the specific issues being addressed. Gap analysis is particularly suitable for sustainable energy management in a particular company but is less suitable for energy management in a particular area. In general, gap analysis includes the following steps.

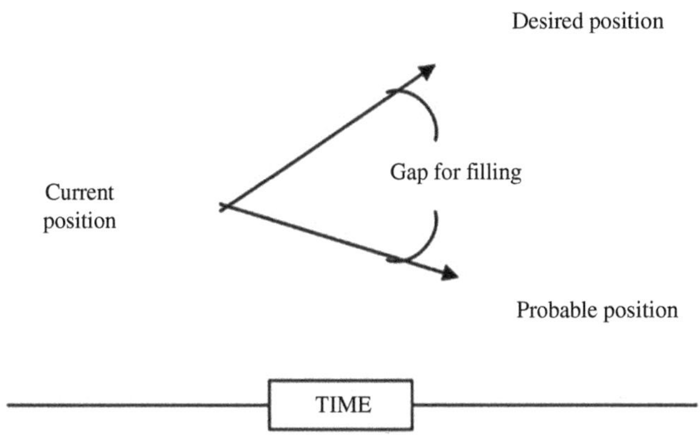

Fig. 3.23 Filling the strategic gap in development.

Determining the current position of the company, based on the estimates of internal indicators. The analysis is performed based on the perception of the current energy status of the company. This stage of analysis can be very comprehensive and complex, especially for companies that use large amounts of energy. It is necessary to determine the following:

- Amount of the total energy consumed in the company for a certain period of time. Quantities should be monitored and expressed at least annually, because in this way certain seasonal fluctuations in consumption can be identified;
- Amount of total energy consumed in certain periods during the year. Depending on the type of activities, significant fluctuations in the quantity of energy consumed are possible;
- Amount of total energy consumed in certain sectors of the company, where it is necessary to separate the amount consumed exclusively for the production of energy from the energy used for administrative and side activities;
- Types (origin) of energy used in the company, as well as the amount of energy from certain sources. In fact, in many cases companies use different types of energy (electricity, heat, coal energy, oil, gas), and it is necessary to accurately monitor the consumption according to certain types of energy resources;
- Amount of energy consumed in the company that was obtained from the company's own resources or alternative sources. There are many technological solutions that enable companies to produce a certain amount of energy themselves, using some of the energy recycling technologies;
- Costs that the company has according to the types of energy resources, i.e., savings achieved by exploiting their own sources of energy;
- Energy costs included in the price of the final product or service;
- Share of energy costs in total costs the company has recorded;
- Energy efficiency of technology to be used, depending on the type of technology, training of employees, the proper management of the work process, control, and many other factors;
- Current emissions in the environment that result from exploitation of energy sources, as well as the efficiency of measures taken in terms of environmental protection;
- Company commitment to change in the field of energy management, which requires compliance with the concept of sustainable development;

- Type, shape, and strength of environmental factors that affect energy management policy, reviewing existing regulations and treaties ratified by the country in which the company or its business unit operates;
- Type and the strength of the impact of consumers who have specific requirements in terms of environmentally responsible and energy-saving operations; and
- Existing requirements of banks, insurance companies, and all other financial institutions with which the company is in contact, and which to some extent evaluate the environmental and energy aspects of the business.

Based on all the above, it is possible to get insight into the three basic parameters that need to be defined in the process of sustainable energy management: energy resources consumption, energy consumption costs, and energy efficiency of business.

Determining where the company is likely to be at a specified time should the current direction and pace of growth and development continue is the next group of data and is based on the extrapolation of goals based on historical data and trends in the past. At this stage, it is necessary to:

- Consider the overall quality of business and the position that the company has had in the past few years. The period for which the analysis is made is determined by the length of operations of the company itself as well as opportunities to get accurate data;
- Consider the energy consumption in previous years, both in terms of type and quantity;
- Determine energy costs that were recorded in the previous period;
- Define the production capacity that has been made on the basis of specific energy consumption;
- Assess the pace of growth expected to continue in the future, which will be based solely on the characteristics of the business. Namely, it is assumed that the company will be developed at the same pace, which means it will use the same amount and type of energy resources, or, if the trend is such, will increase or decrease the amount of energy consumed depending solely on the capacity of production to be achieved. These estimates are based only on the business policy of the company, which is usually market conditioned;
- Estimate the amount and type of energy the company will spend in the next period, without any special changes or adjustments, neglecting the needs for sustainable energy management;

- Evaluate the energy costs that will be recorded if the company continues developing at the desired pace;
- Estimate the future share of the cost of energy resources used in the unit price of the final product;
- Estimate the possibilities for providing the desired types and quantities of energy in the future; and
- Estimate the future emissions in the environment that will be created by energy resource consumption based exclusively on the economic principles of the business.

The basic data to be obtained by this analysis include the estimation of the amount of future energy consumption, estimation of future costs of energy consumption, and assessment of future energy efficiency. The estimation of the condition in which the company will probably find itself is a process related to a certain degree of uncertainty, since it depends primarily on the stability of general business conditions in the country. Also, the precision of this estimation can be considered acceptable for a period of 5–10 years; however, in countries with unstable business conditions, this period may be shorter.

Determining the position in which the company wants to be with the use of sustainable energy management, based on desired goals. The company that wants to become and remain a part of the modern world market has to adapt its business to the requirements of sustainable development, which means that, among other things, it must incorporate in its development plans sustainable energy management. Recognizing this, companies can assess their own future after the implementation of sustainable energy management measures based on the following:

- Estimation of future total energy consumption based on constant savings, rationalization, and optimization;
- Estimation of future costs of energy incurred as a result of reduced consumption of energy or energy-efficient operations;
- Estimation of investments required to achieve savings in energy consumption, which involves estimating the size of the investment and determining the economic and noneconomic effects of investments in this area;
- Estimation of future levels of energy efficiency that can occur as a result of the implementation of sustainable energy management measures;
- Definition of the share of energy consumed from its own or other renewable sources in the total amount of energy consumed in the company;

- Estimation of financial and nonfinancial effects that may occur by reducing harmful emissions into the environment;
- Estimation of the condition and reaction of the environment to environmentally responsible and irresponsible behavior of the company;
- Estimation of compliance of energy-efficient operations of the company with the requirements of legislation, both domestically and globally; and
- Estimation of the situation in which the company wants to be if it applies sustainable energy management, which is a particularly sensitive category in the process of forecasting in general.

The future is in essence unpredictable, and there is always a degree of risk. On the one hand, companies are faced with the trend of environmentally responsible business in general, and on the other, there is a clear awareness on the existing energy resource limitations. All this points to the need to develop specific tools that will enable more effective prediction in this field.

Today, there are very effective criteria for assessing economic effects and justification of investments, but currently there are no mechanisms by which the effects of a long-term implementation of sustainable energy management could be predicted with great precision. Companies will introduce it into their business because it is a requirement of sustainable development and an imperative of civilized business, but predicting its effects over the long term is currently at the level of assumptions. It is necessary to distinguish between an unstable and stable environment and adjust the techniques underlying the control of the strategy. If the environment is stable, then it is relatively easy to predict, and the role of control is to observe the progress of companies. In an unstable environment, when prediction is difficult, the task of control is to assess the assumptions, goals, and components of the strategy.

Determining the size of the gap between the extrapolated and projected goals includes determining the difference that will occur in future business if certain changes are or are not introduced. Any change in itself is a process that has its causes and effects, including changes to sustainable energy management:

- The gap will be greater if the company deals with the activities connected with high energy consumption;
- The gap will be greater if the company uses only nonrenewable energy sources;
- The gap will be greater if the company recorded low energy efficiency indicators; and

- The gap will be greater if the company is planning big changes in terms of transitioning to a sustainable way of managing energy.

The whole process of gap analysis is linked to a number of assumptions, and defining the gap is the last step of the method and is often impossible to define precisely. Properly conducted analysis of the gap that exists between the state that would be realized without introducing changes and the one that can be achieved by using sustainable energy management must provide the following information:

- The total change in the amount of energy spent in the company, since the process of sustainable energy management results in a decrease in the amount of energy consumed;
- The changes in the quantity of energy consumed, depending on the type of energy, because it may happen that the implementation of the process of sustainable management of energy reduces consumption of only one type of energy resource, which usually depends on the specifics of the technological process;
- The size of the change in energy efficiency, since the process of sustainable energy management leads to reduced energy consumption for producing a product unit; and
- The financial impact of implemented changes, which are related to showing changes in revenues and expenditures that inevitably occur in the process of implementation of the sustainable management of energy, by which increased costs in the form of investments in energy-efficient operations can be expected. Savings will later be registered in terms of positive effects of implemented investments.

From the point of view of proper assessment of economic efficiency of the system of sustainable energy management, it is particularly important to consider financial aspects and the time in which the above effects can be expected. In this way, the company will be able to predict in the long term and implement the necessary allocation of resources and anticipate the moment in which to expect the first positive financial effects. In this way, financial instability can be avoided or significantly reduced.

Establishing programs to fill the gap involves a whole series of measures and activities directly related to the need to surpass the instability that arises as a consequence of all business transactions, which can be very strong if the sustainable energy management is introduced in companies whose business is connected with large energy consumption. At this point, it is necessary to predict the following:

- Time frame required to move from traditional to sustainable energy management, which must be coordinated and integrated with other management plans (usually a minimum period of 5 years is required);
- Financial resources required for implementation of the measures and activities that lead to full implementation of the process of sustainable management of energy. When using internal sources, the company must anticipate the intensity of the financial effects that may occur by investing in this aspect and neglecting other aspects of business. If the financial resources are provided from external sources, it is necessary to use adequate methods to predict expected cash flow and current liquidity problems, as well as measures to solve these problems; and
- Changes in the sphere of human resources, which must be determined both by the number and quality. The implementation of sustainable energy management is associated with changes in traditional business, which sees energy solely as an expense. If the company is able to pay for it, there are no obvious reasons to make any changes. Sustainable energy management looks at energy as a resource that should be spent optimally, regardless of the company's financial strength. It is necessary to provide training for employees at various levels, with possible involvement of people who are specially trained for the introduction, implementation, and monitoring of sustainable energy management.

At this point, the company is faced with the need to cover changes that are imminent or the gap that is identified by planning and management in the sphere of its three basic resources: time, money, and people. For these resources, adequate analysis, development plans, and ways of monitoring and control are necessary because it is essential that these three resources effectively support all activities required for the complex operations of transition to a sustainable energy management system, as shown in Fig. 3.24.

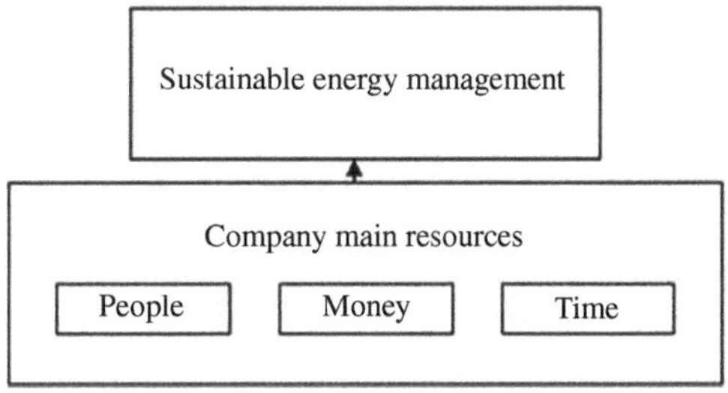

Fig. 3.24 Basic resources of a company in the function of sustainable management of energy.

Determining the mechanisms of control, or a set of control points to meet the gap. The introduction of sustainable energy management is a big, strategic change in the way of thinking and operating and must therefore be regarded as a kind of strategic management process. The process of strategic management involves iteratively connected steps that are aimed at ensuring efficient and effective business. On the other hand, a strategic decision-making process usually requires long-term engagement of the company, during which the company changes the way it operates. Accordingly, assessment, control, and auditing are needed to determine whether the strategy as a product of strategic management is in accordance with, above, or below expectations and planning. For these reasons, the audit itself should involve several areas including:

- Evaluation of internal harmonization of strategic goals with other goals of the company, as a result of the need to harmonize changes in the field of energy management with other plans of the company, which should all be integrated together into a single management system. The goals of business are numerous, but they all lead to opportunities for income generation, by which the company fulfills its mission and the purpose of its existence. On the other hand, modern companies determine social responsibility as one of their goals that promotes business in accordance with the concept of sustainable development,
- Evaluation of the quality of the process of analysis, which is the basis for the development of the strategy, including a realistic assessment of the goals set as desired or likely in the process of gap analysis, and according to which further assessments and management plans are made. At the time of setting goals, their supposed value has been set as well as their size or some other characteristics, and after some time, it is possible to validate a company's ability to achieve certain predictions. Modern business is increasingly associated with predictions rather than analysis of previous events, so attention has to be paid to the selection of the required methods for predicting;
- Evaluation of the very content of the strategy (direction, pace, and method of growth), which was selected as the content for the gap analysis. Depending on the goals and the size of the estimated gap, the company chooses its own way for the implementation of sustainable energy management. The above process can be implemented in whole or in part, faster or slower, with fewer or more costs, but requires monitoring of changes, as well as other enterprise business strategies that exist in the given period of time;

- Evaluation of the company's opportunities to achieve the formulated strategy requires a remarkable ability to assess if the company is capable of filling the gap and reaching the set goals, after the internal and external analysis, or after setting goals and defining the gap. Sustainable energy management is often a desired goal, but its implementation is associated with a number of changes and investments, so most companies are not able to immediately implement a complete process of sustainable energy management. The gradual transition to a new way of energy management through the implementation of smaller, easily achievable goals is the safest way for most companies, although it takes more time; and
- Evaluation of results brought about by the implementation of the strategy is a necessary step in assessing the effectiveness of the method of gap analysis. The gap analysis method is based on results, and its main advantage is that it provides the opportunity to precisely measure and assess whether the use of energy management has led to the desired goals. If so, the company gets valuable experience, and if goals are not achieved, or are not fully realized, the company is able to find out why it happened and to identify where the error occurred. In this way, it can correct its own management plan. The most common errors that occur in the process of transition to sustainable energy management are due to misjudgment of abilities and nonobjective setting of goals.

Gap analysis is based heavily on projections and assumptions. It is extremely demanding in terms of a company's ability to assess its own abilities. Most companies have a positive attitude toward the implementation of sustainable energy management, but the first problems arise even during the internal analysis of capabilities to implement something like it in practice. If the problems are ignored and companies overestimate their own strength, their goals will be too high and a gap will be present, which can cause the company to give up the sustainable management of energy because it may seem unachievable, or the company will engage in a long period of changes based on nonobjective assumptions.

For the reasons stated above, companies need to understand the complexity of the process of introducing sustainable management of energy and observe it as a big change that requires a very serious approach. The application of available methods and management techniques will reduce the possibility of error and increase the likelihood of successful implementation. No company achieves all its goals, but a modern company should not be resistant to changes in the field of energy management. The methods for implementing sustainable energy management are all specific, and there are several implementation methods to choose from. The factor that in most cases affects the choice of a particular method is the activity that the company deals with, but above all, its efficiency and experience in managing change, i.e., the transition from traditional to sustainable energy management.

Case study

Evaluation of corporate energy management practices of energy intensive industries in Turkey, Energy, Volume 45, Issue 1, 2012, Pages 81–91, Ates et al. https://www.sciencedirect.com/science/article/abs/pii/S036054421200223X.

References

Amorim, E. V. (2014). Sustainable energy action plans: Project management intercomparison. *Procedia Technology*, *16*, 1183–1189.

Antony, R., & Govindarajan, V. (2017). *Management control system (SIE)* (12th ed.). McGraw Hill Education.

Armstrong, M. (2001). *A handbook of management techniques: The best-selling guide to modern management methods* (3rd ed.). Kogan Page Publishers.

Barletta, I., Despeisse, M., Hoffenson, S., & Johansson, B. (2021). Organizational sustainability readiness: A model and assessment tool for manufacturing companies. *Journal of Cleaner Production*, *284*, 125404.

Beisheim, B., Krämer, S., & Engell, S. (2020). Hierarchical aggregation of energy performance indicators in continuous production processes. *Applied Energy*, *264*, 114709.

Bhattacharjee, R. M., Dash, A. K., & Oaul, P. S. (2020). A root cause failure analysis of coal dust explosion disaster—Gaps and lessons learnt. *Engineering Failure Analysis*, *111*, 104229.

Bonnardel, N., & Didier, J. (2020). Brainstorming variants to favor creative design. *Applied Economics*, *83*, 102987.

Chen, C. F., Feng, K. L., & Ma, H. (2020). Uncover the interdependent environmental impacts associated with the water-energy-food nexus under resource management strategies. *Resources, Conservation and Recycling*, *160*, 104909.

Comoglio, C., & Botta, S. (2012). The use of indicators and the role of environmental management systems for environmental performances improvement: A survey on ISO 14001 certified companies in the automotive sector. *Journal of Cleaner Production*, *20*, 92–102.

Duke, A. (2020). *How to decide: Simple tools for making better choices*. Portfolio (Illustrated ed.).

Fleacă, E., & Fleacă, B. (2016). The business process management map—An effective means for managing the enterprise value chain. *Procedia Technology*, *22*, 954–960.

Goel, S. L. (2007). *Management techniques: Principles and practices*. Deep & Deep Publications.

Gopinath, R., Kumar, M., Joshua, C. P. C., & Srinivas, K. (2020). Energy management using non-intrusive load monitoring techniques—State-of-the-art and future research directions. *Sustainable Cities and Society*, *62*, 102411.

Haley, B., Gaede, J., Winfield, M., & Love, P. (2020). From utility demand side management to low-carbon transitions: Opportunities and challenges for energy efficiency governance in a new era. *Energy Research & Social Science, 59*, 101312.
Haraguchi, M., Siddiqui, A., & Narayanamurti, V. (2019). Stochastic cost-benefit analysis of urban waste-to-energy systems. *Journal of Cleaner Production, 224*, 755–765.
Horowitz, I., & Woo, C. K. (2006). Designing Pareto-superior demand-response rate options. *Energy, 31*, 1040–1051.
Huffmire, D. W., & Holmes, J. D. (2006). *Handbook of effective management: How to manage or supervise strategically*. Libraries Unlimited.
Johnstone, L., & Hallberg, P. (2020). ISO 14001 adoption and environmental performance in small to medium sized enterprises. *Journal of Environmental Management, 266*, 110592.
Kochhar, S. (2012). *MS-43 management control systems* (2019th ed.). Gullybaba Publishing House (P) Ltd.
Krampell, M., Solís-Marcos, I., & Hjälmdahl, M. (2020). Driving automation state-of-mind: Using training to instigate rapid mental model development. *Applied Ergonomics, 83*, 102986.
Locmelis, K., Blumberga, A., Bariss, U., & Blumberga, D. (2017). Energy policy for energy intensive manufacturing companies and its impact on energy efficiency improvements. System dynamics approach. *Energy Procedia, 128*, 10–16.
Lue, R. W., Byars, L. L., & Ibrahim, N. A. (2015). *Human resource management*. McGraw Hill.
Luo, T., Wu, C., & Duan, L. (2018). Fishbone diagram and risk matrix analysis method and its application in safety assessment of natural gas spherical tank. *Journal of Cleaner Production, 174*, 296–304.
Martins, F., & Fonseca, L. (2018). Comparison between eco-management and audit scheme and ISO 14001:2015. *Energy Procedia, 153*, 450–454.
McGlynn, R. P., McGurk, D., Effland, V. S., Johll, N. L., & Harding, D. J. (2004). Brainstorming and task performance in groups constrained by evidence. *Organizational Behavior and Human Decision Processes, 93*, 75–87.
McGuire, W. (2014). The effect of ISO 14001 on environmental regulatory compliance in China. *Ecological Economics, 105*, 254–264.
Min, Z., Morgenstern, P., & Marjanovic-Halburd, L. J. (2016). Facilities management added value in closing the energy performance gap. *International Journal of Sustainable Built Environment, 5*, 197–209.
Murmura, F., Libertadore, L., Bravi, L., & Casolani, N. (2018). Evaluation of Italian companies' perception about ISO 14001 and eco management and audit scheme III: Motivations, benefits and barriers. *Journal of Cleaner Production, 174*, 691–700.
Namany, S., Al-Ansari, T., & Govindan, R. (2019). Sustainable energy, water and food nexus systems: A focused review of decision-making tools for efficient resource management and governance. *Journal of Cleaner Production, 225*, 610–626.
Ogbeibu, S., Emelifeonwu, J., Senadjki, A., Gaskin, J., & Kaivo-oja, J. (2020). Technological turbulence and greening of team creativity, product innovation, and human resource management: Implications for sustainability. *Journal of Cleaner Production, 244*, 1108703.
Orcos, R., Pérez-Aradros, B., & Blind, K. (2018). Why does the diffusion of environmental management standards differ across countries? The role of formal and informal institutions in the adoption of ISO 14001. *Journal of World Business, 53*, 850–861.
Rabaia, M. K. H., Ali Abdelkareem, M., Sayed, E. T., Elsaid, K., Chae, K. J., Wilberforce, T., et al. (2021). Environmental impacts of solar energy systems: A review. *Science of the Total Environment, 754*, 141989.
Smith, K., Bastin, M. E., Cox, S. R., Valdés Hernández, M. C., Wiserman, S., Escudero, J., et al. (2019). Hierarchical complexity of the adult human structural connectome. *NeuroImage, 191*, 205–215.

Thakkar, A., Chaudhari, K., & Shah, M. (2021). Comprehensive survey on energy-efficient power management techniques. *Procedia Computer Science*, *167*, 1189–1199.

Tsani, S., & Menegaki, A. N. (2018). The energy-growth Nexus (EGN) checklist for authors. In *The economics and econometrics of the energy-growth Nexus* (pp. 377–380). Academic Press (chapter 11).

Waqas Iqbal, M., & Kang, Y. (2020). Waste-to-energy supply chain management with energy feasibility condition. *Journal of Cleaner Production*, 125231.

Whitehurst, J., & Hamel, G. (2015). *The open organization: Igniting passion and performance*. Harvard Business Review Press.

Wu, G., Sun, C., & Xu, S. (2019). Source control or end-of-pipe control: Mitigating air pollution at the regional level from the perspective of the total factor productivity change decomposition. *Energy Policy*, *129*, 1227–1239.

Zhang, Y., Wei, Y., & Zhou, G. (2018). Promoting firms' energy-saving behavior: The role of institutional pressures, top management support and financial slack. *Energy Policy*, *115*, 230–238.

Zhou, L., & Dai, Y. (2020). Which is more effective in China? How communication tools influence public acceptance of nuclear power energy. *Energy Policy*, *147*, 111887.

Ziębik, A., & Gładysz, P. (2018). Systems approach to energy and exergy analyses. *Energy*, *165*, 396–407.

CHAPTER 4

Planning of energy management

Contents

Traditional concept of energy management

Energy management is in most cases planned in theory and implemented in practice in a traditional way by implementing the already established principles of management that can be applied to almost any other human activity. Classical management science has developed valuable and effective mechanisms for management in all sectors of human labor, in which, depending on the needs and specific nature of work, certain adjustments can be made but are predetermined and largely known in advance. Thus, developed management, based on the learning of James A.F. Stoner, Ishak Adizes, Peter F. Dracker, and other founders of modern management science, has greatly improved business in the second half of the 20th century and increased efficiency and effectiveness of human labor in all areas where it has been applied (Stoner, Freeman, Gilbert, & Nambudiri, 2018).

Traditional management has been adapted to modern business operations under the conditions of globalization and controlled flows of people, goods, and capital, and it greatly enables the persistence of certain system and socioeconomic relations in the world. With changes in social consciousness and the first criticisms of the existing relations in the global economic and political scene, there emerges the need for certain changes in many spheres of interest for modern mankind. This imposes the need to make changes and

Sustainable Energy Management
https://doi.org/10.1016/B978-0-12-821086-4.00008-8

adjustments in the field of management that would empower this science to respond to challenges of doing business in the future (Robbins, Judge, & Vohra, 2016).

Energy management is a particularly demanding challenge in all areas, especially in terms of the need to manage energy in a manner required by the modern world, which is not a simple application of the process of management in the field of energy. With the increasing intensity of energy use in the world, there has been a need to develop adequate energy management at all levels, where it is necessary to develop mechanisms that will react and adapt the whole process of management, depending on the character and intensity of changes taking place in the environment.

The process of energy management can be seen in an extremely simplified way as a process that takes place from the moment of extraction of energy raw material to its final consumption and disposal of the waste energy. This process takes place in an extremely complex environment, is subject to the impact of numerous factors, and represents certain challenges in countries and regions where very stable conditions prevail and parameter changes are reduced to minimum. A particular problem is power management in the areas where frequent changes occur regarding type and intensity. Moreover, particularly challenging is the attempt to develop a unique model for energy management globally.

With the introduction of the principle of sustainability, as the only principle of the development of mankind in the future, the issues of energy management are given a new dimension, which greatly differs from traditional management and imposes the need for the adoption of new solutions. In the most widely accepted context, energy management involves the use of a sustainable model of strategic management in the field of monitoring of energy flows in terms of production, transport, transformation, and consumption.

Given the complexity of energy management, it is necessary to use the available approaches and modify them in accordance with the requirements of sustainability. Creating an adequate model of sustainable energy management is only possible with the integration of sustainable development requirements, particularities of the energy sector, and the science of strategic management. By integrating these criteria, a new interdisciplinary approach to energy management has been created that has its own characteristics and is subject to changes depending on many factors. Therefore, it is primarily necessary to give a brief overview of the development of the science of strategy and strategic management in general.

A single definition that would include all the features and aspects of the strategy would be long and complicated, so each author who contributes to development of modern management emphasizes one or more aspects.

Perhaps the earliest definition of the strategy originates from the ancient Greek writer Xenophon: "The strategy is the knowledge of the business which you propose to be achieved" (Radovanović (Golušin), Dodić, & Popov, 2013). Xenophon clearly indicates the direct responsibility of those who manage (managers) and emphasizes the link between leadership and strategy formulation.

The important contributors to management science, Igor Ansoff and Anthony McDonnell point out that strategic management is a systematic approach to managing change (Ansoff, 1979). By this definition the authors make a distinction between strategy and policy, where the policy is a universal decision that is always made in the same way if the same circumstances occur, while the strategy applies similar principles, allowing for different decisions to be made depending on the concrete circumstances.

On the other hand, Ishak Adizes points out the following: "Instead of the story of a manager who plans, organizes and so on, we should speak of the management team that performs these functions. The roles of producers, managers, entrepreneurs and integrators must be performed by the complementary team, because no single person can be responsible for all of them" (Adizes, 2018).

Kenichi Ohmae's definition of strategy explicitly emphasizes the importance of meeting the expectations of customers' needs as the driving force of the strategy and points out the competitive aspect of strategy. Ohmae defines strategy as follows: "The way in which corporation seeks to differentiate itself in a positive way compared to its competitors, using its relevant forces to better meet customers' needs" (Ohmae, 1982).

In the definition by Kenneth Andrews, strategy is seen as the relationship between the goal and possibilities of the realization of that goal, making the strategy sensitive to values and culture as much as to chances and opportunities arising in the environment. Andrews gives a broad definition of strategy: "Modeling of the main aims, purposes and achievements and the important policies or plans to achieve these goals, which are stated in such a way that they define in which business the company is or it will be, as well as the current type of company or what the company should be" (Andrews, 1987).

Authors Samuel Certo and Paul Peter point out that strategic management is a continuous, iterative process targeted at the maintenance of organization completely appropriate to respond to its environment. There are several traditional scientific approaches to strategic management, and all are in use in all areas of human activity around the world. Authors Certo and Peter define the process of strategic management as a continuous, iterative process that involves several steps (Certo & Peter, 1989):

1. Analysis of the environment.
2. Directing the organization (mission and objectives).
3. Formulation of strategy.
4. Implementation of strategy.
5. Strategic control.

The process of strategic management in this case is defined as the process that is going on in an infinite number of repetitions, as shown in Fig. 4.1.

Thus, the defined process of strategic management has been widely accepted in practice because of its simplicity. Complexity of a particular process does not necessarily lead to its improvement, but such a simplified approach to strategic management is not quite acceptable for energy management, which must respect the requirements of sustainability. The main objection to the presented model is primarily the narrow scope of defining the strategy, i.e., the ways to achieve the desired goals. In addition, the process of control does not take into consideration legal limits but is only based on examination of the achieved objectives, which are often not properly set, measurable, or given in an exact way.

Authors Gerry Johnson, Kevan Scholes, and Richard Whittington developed a similar and yet redefined model, because the process of strategic management, which could serve as the starting point for developing the model of sustainable management of energy, is observed in three steps. According to these authors, the process of strategic management consists of the following (Johnson, Scholes, & Whittington, 2008).

1. Strategic analysis, which consists of analysis of the environment, analysis of the expectation of goals and strengths, and analysis of the resources.
2. Strategic choice, which consists of generating options, estimation of options, and selection of strategies.

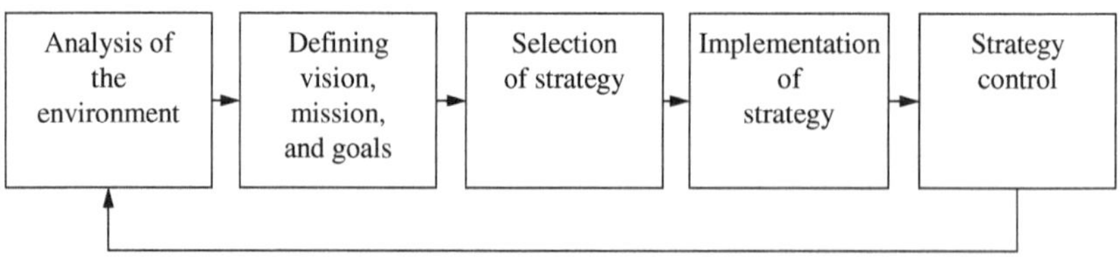

Fig. 4.1 The process of strategic management according to Certo and Peter.

3. Strategic implementation, which consists of planning resources, organizational structure, and people and systems.

Contrary to the previously shown and proposed basic model, the process of strategic management of energy is based on the relationship between the three basic steps, which can be summarized in Fig. 4.2.

The model presented above suggests to some extent the application of new solutions that can be regarded as acceptable for the sustainable management of energy, primarily because of the emphasis on the connection of the three basic elements, especially implying a particular framework for strategic choice. Namely, the process of energy management is very complex, and it is therefore necessary that in the planning stage there is the possibility of defining and redefining the choice of strategy to be used for the realization of the set goals. The basic elements of this model presented here have their own characteristics and subsystems.

Strategic analysis includes processes of analysis and prediction of environment, analysis of the resource capabilities, and expectations and the strength of the main social groups. These analyses should result in understanding of the strategic position of the region, opportunities and threats, and definition of the main goals of the process of energy management (Sapmpler, 2021).

Strategic choice also consists of three elements. It includes identification, description, and evaluation of strategic options and choice of strategy that could be implemented in the process of energy management. According to the theory and science of management, the determination for further actions should be defined at this point, i.e., the selection of several options: the stabilization of energy production and consumption, expansion, reduction or pause, and a more or less radical strategic shift in which the process of energy management has to undergo significant changes. Respecting the very concept of sustainable management of energy, the strategic approach to

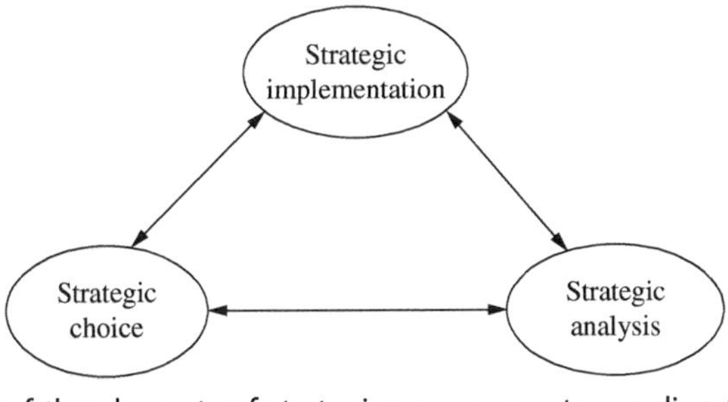

Fig. 4.2 Model of the elements of strategic management according to Johnson and Scholes.

management can be considered more acceptable than the one previously discussed because it opens the possibility for more strategic choices and allows the implementation of the strategic shift as an option that takes into account the need for significant changes in dealing with energy resources and energy consumption (Elgersma, 2017).

Strategic implementation refers to issues that determine the fulfillment of requirements for the implementation of selected strategic options. For this reason, it includes planning input, defining the appropriate organizational structure and selection of management style, and corporate culture ánd recruitment.

Authors Higgins and Vincze have a similar approach, and in the process of strategic management, they start from formulation of the vision, mission, and developmental goals and preparation for implementation (Vincze & Higgins, 1992).

1. Formulation of vision, mission, and goals.
2. Setting strategic goals.
3. Development of strategy.
4. Implementation of the selected strategy.
5. Assessment and control of strategy.

The presented model of strategic management requires greater efforts in the field of planning, which has been realized in the first two strategic steps, when the vision and mission have been defined. Particular attention has been paid to setting strategic goals. Although one step has been divided into two steps, the model has to be defined and analyzed as a separate model of management, as shown in Fig. 4.3.

The presented model of strategic management requires greater efforts in the field of planning, which has been realized in the first two strategic steps, when the vision and mission have been defined. Particular attention has been paid to setting strategic goals.

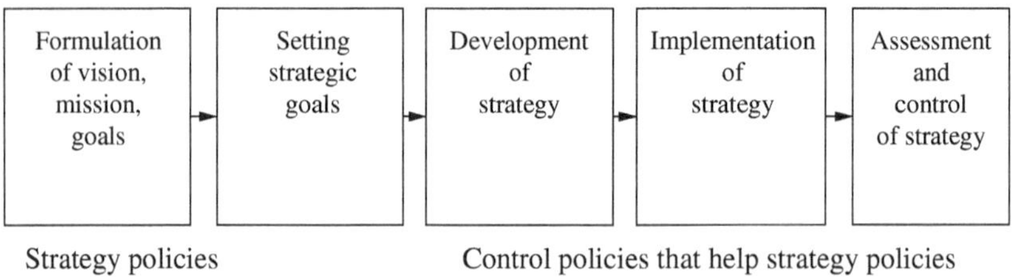

Fig. 4.3 Model of the process of strategic management according to Higgins and Vincze.

Therefore, the wrong definition of the strategic goals in the early stage of planning can have far-reaching good and bad consequences in terms of implementation and success of the entire strategy. This is also the greatest weakness of the proposed model, since the process of sustainable energy management is a process that has consequences of a proper or improper planning and definition of the objectives will be felt after many decades, which questions the purpose of the control.

Taking into account all the strengths and weaknesses of the described strategic models, we should present the model that can be regarded as the most appropriate one for the initial creation of the model of sustainable development of energy, which consists of eight strategic steps (David, David, & David, 2020):

1. *Defining vision and mission of the company*, defining the main attitudes that direct the attitudes of the company toward the future (defining vision), and determining the role of the company in a given system of doing business (defining mission).
2. *Analyzing and forecasting the direct and indirect environment* should provide (a) diagnosis of the current strategic position of the company and (b) assess opportunities and threats that will be created with the projected development of the environmental factors.
3. *Analyzing and forecasting internal possibilities* of the company and the conditions for their provision.
4. *Defining objectives* through quantitative and qualitative determination of the type of business performance the company strives for and in relation to that it measures its effectiveness and efficiency in the implementation of its mission and strategy of the company.
5. *Identifying the critical factors of business success* and determination of activities that a company must undertake to operate in the selected field of activities. Then concrete critical factors are searched for so as to gain competitive and comparative advantages.
6. *Identifying and evaluating strategic options* (alternatives) and choice of appropriate strategies, i.e., searching for potential ways and instruments for the selection of goals and selection of the business strategy.
7. *Implementation*, i.e., creation of the appropriate plan of business culture and management style, the plan to identify resources, the selection of the organizational structure, systems of motivation, and estimation of the value of the work so as to implement the determined business strategy.

8. *Monitoring and auditing*, i.e., measurement and evaluation of the plan (control) to know whether the plans have been achieved in accordance with the possibilities of companies (audit).

The proposed method of management consists of several more steps compared with the previous ones, which to some extent complicates its implementation, but on the other hand, provides an opportunity for detailed planning and monitoring, which is imperative in the process of sustainable energy management. This process is influenced by a number of factors, lasts long, and has no precisely defined end. Additionally, its implementation can significantly influence economic, ecological, and any other development in general. The process of strategic management could in this case be as presented in Fig. 4.4.

The proposed initial model has its advantages and disadvantages. The main disadvantage of the proposed model is its complexity, which can lead

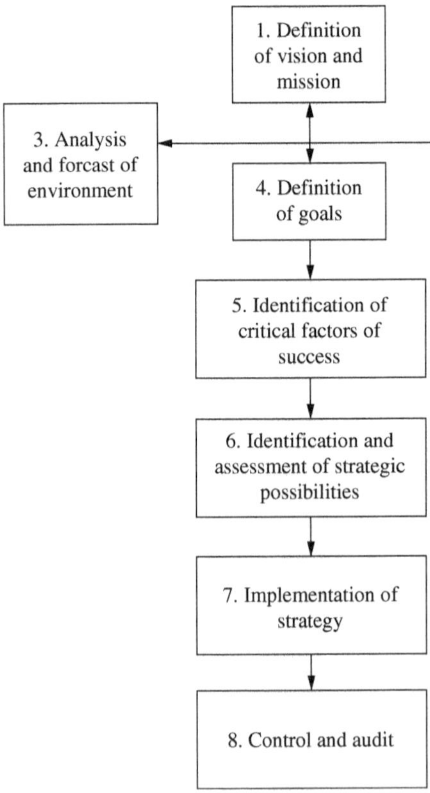

Fig. 4.4 Elements of the process of strategic management as a starting point in defining the model of sustainable energy management.

to problems in the planning of resources, people, and time. The main advantage of the initial model is the need to more carefully plan all the steps of management, because the implementation of the model is in this case the last, eighth step. All other stages involve preparation with particular emphasis on the analysis of the internal and external environment. For the purposes of energy management analysis of the environment this is of primary and probably the greatest importance, because it is very complex and multidimensional. It requires the ability to estimate and forecast in spheres that are not closely related to the sector of production and consumption of energy (Capehart, Turner, & Kennedy, 2016).

Each company should strive to fit, in accordance with their abilities and goals, into the above strategy of sustainable energy development and adjust their management according to it. However, all business entities are not and cannot at the same time fully implement the complete process of sustainable energy management. Modern management science recognizes this fact and offers four basic attitudes a company can adopt so as to implement the process of energy management, as follows (Smith & Parmenter, 2015):

The legal position, which means the company would voluntarily and without question apply all the laws, rules, and regulations in the field of sustainable energy management, but try to apply these rules to their advantage by making the necessary changes and innovations in their business operations;

The market position, as acceptable for companies that would base their operations on meeting the desires of their customers that are related to sustainable energy management;

The attitude of all stakeholders, which extends the above attitude because, apart from customers, it meets the requirements of other interested parties;

The dark green attitude, which promotes life that should be lived in greater harmony with nature. This position is currently very difficult to apply at the current level of awareness and business.

By choosing the proper attitude, a company can adequately plan its future energy development. Each individual company has to review its own goals and approach problem-solving in ways that are estimated as optimal. Respecting the regulations in the field of environment protection, energy management, and sustainable development, as well as accepting ecological management as a business strategy, it is possible to actively participate in the process of environment protection and the concept of sustainable development as a whole (Adizes, 2018).

From the standpoint of a particular company, there are three main ways in which company management can act to solve its problems in the field of sustainable energy management (Rossa, Doty, & Turner, 2018):

- *Inactive management*, in which there are no activities in terms of understanding problems of energy management, so that there are no activities in terms of their solving and prevention;
- *Reactive management*, which reflects the situation in most companies today and implies (possible) problem-solving only when problems occur. In such companies, no preventive measures have been applied, and certain actions in terms of sustainable development of energy are planned only when the company is forced to, and even then, that process is usually slow and sporadic;
- *Proactive management*, which is imperative of any modern energy-responsible business, where all activities are directed toward planning and prevention. The knowledge of the specifics of sustainable energy management and the considerable efforts toward its implementation are built into the business policy of the company and become its integral part.

Selection of the approach in the field of energy management is conditioned by the influence of a series of internal and external factors that define each company, as well as the social and economic circumstances in which the company operates. On the other hand, planning of energy management in terms of a particular state or region is connected with an even greater number of factors that determine it at a certain point and that are changing very frequently. Therefore, the development of an adaptive and yet clear model of sustainable energy management is crucial.

Sustainable approach to energy management

Previous experience, theory, and practice clearly indicate the need to develop special mechanisms aimed at proper management of energy in all stages and at all levels in all timeframes. The need for energy management has existed since ancient times in various forms, but only modern science and management practices, respecting the principles of sustainable development, have provided an adequate framework for the realization of sustainable energy management.

When designing sustainable energy management, it is necessary to start from the broadest framework that defines processes in this area. This relates primarily to two main factors that are extremely important: the size of the strategic issue that needs to be addressed and the estimated time required to implement the proposed strategy of energy management. A sustainable approach to energy management has been created because of the two characteristics that determine it more than numerous changes that mankind has encountered through history.

The *time dimension* of energy issues is particularly emphasized and greatly determines the ways of energy management, as well as consequences that may result from it. Meeting the demand for energy has existed throughout the history of modern mankind, and needs have increased as energy resources have been exploited faster and in a more efficient way, but the situation at the end of the 20th century led to the need to reconsider the trend of uncontrolled energy consumption, to slow down and redefine it in the ways that would be in accordance with the concept of sustainable development.

The future of energy management is in a certain way defined by a clear determination to slow down the exploitation of nonrenewable resources, to use energy efficiently, and to raise awareness of all energy users that they have to adjust to the new way of doing business and thinking.

These fundamental changes in energy management are extremely complex, comprehensive, and linked with a number of smaller and larger changes in all spheres of human activities. It is necessary to redefine the existing legislation, adopt appropriate strategies at global and national levels, and adjust business operations of each business entity and individual consumer.

Apart from the time required to use sustainable energy management strategies, in order to select the type of strategic implementation, it is necessary to consider the size of the strategic problem. The energy problem is global and is characterized by a number of factors primarily related to the increase in global needs for energy, uneven distribution of energy resources, and high degrees of dependence of people on the energy they have at their disposal. Problems with energy lead to minor or major problems in economy, society, relations between countries, and relations in the global community. It is expected that these problems will increase and become more complex in the future. It is therefore quite correct to assume that the energy problem is one of the largest and most complex strategic problems of today.

When a company is faced with minor problems (small energy consumers) and there is adequate time available, the implementation of the strategy can include evolutionary or gradual (incremental) changes. If the problems a company encounters are minor, but the period of time is short in which it is necessary to solve the problems, which is often the case when new legal regulations are adopted that sharply define changes in energy consumption, activities are done by management (managerial intervention) directly where the problems have arisen.

If the company is a big producer or consumer of energy, it is necessary to carefully select the method of implementation of strategy of the sustainable management of energy. Depending on how much time the organization has to solve the problem, there are two possible strategies. If the company has a long period of time, and the problems that emerged are big, sequential intervention, i.e., intervention in stages, is required. However, if the problems cannot wait to be solved, complex intervention is applied, and the management of the company has to synchronize (adjust) the changes in all parts of the company (Fig. 4.5).

Changes in national energy management largely depend on the resources the country possesses and the way in which energy is being spent. The strategy of sustainable development defines energy problems in general as those that require too much time to be solved, i.e., longer than the lifetime of the current generation. The energy problem will be a part of mankind for an indefinite period of time in the future.

The selected strategy of sustainable energy management is a highly complex process influenced by many factors. On the other hand, regardless of the clear definition of the proposed models for energy management, it is often not possible to select and implement a certain strategy without intermediate changes and adjustments. In addition, changes in the energy sphere are numerous and diverse, and their impact on the sustainable management of energy is often associated with a series of economic and political changes. Because of these issues, there are difficulties in clearly defining what sustainable energy management actually is. Every possible definition can be redefined and changed depending on many factors.

		Timeframe for the change of implementation: Long	Timeframe for the change of implementation: Short
The scope of strategic problem	Big	Sequential interventions	Complex interventions
	Small	Evolutionary interventions	Managerial interventions

Fig. 4.5 The types of possible implementations of the strategy of sustainable management of energy.

Energy management can be defined as the process of planning, directing, implementing, and controlling the process of generation, transmission, and consumption of energy. Energy management is a certain synthesis of energy and the concept of modern management, i.e., the application of modern hypotheses of management in the energy sector. Apart from that, in stating the main hypotheses in terms of energy management, modern management starts with the assumptions that it is possible to preserve and keep energy stability for current and future generations. Therefore, modern energy management can be seen as a kind of synthesis of the three exact sciences: energy industry, sustainable development, and management, which are interconnected and mutually conditioned, as shown in Fig. 4.6.

Sustainable energy management is a new concept, idea, and approach that requires many changes in the traditional way of understanding and interpreting energy management at all levels. Sustainable energy management integrates many features of the environment and cannot therefore be interpreted as a concept adopted and defined once and for all, but must be constantly modified and adjusted in accordance with changes in the three areas that define it, and in accordance with the specific country or region where it has been applied. Basically, the concept of sustainable energy management is defined by the following parameters:

- Management, i.e., accepted theories, existing experience, state regulations, company orientation, the requirements of all stakeholders, degree of education, awareness, and commitment as well as general orientation toward socially responsible business;
- Energy industry, legislation, existing energy resources, new energy resources, energy efficiency level, the level of technology development,

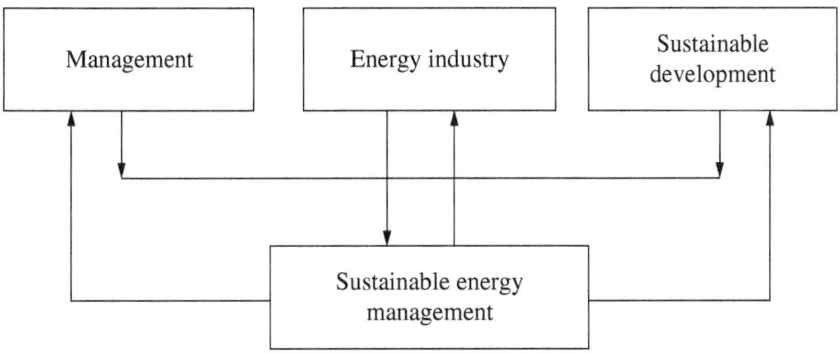

Fig. 4.6 The main concepts of the sustainable management of energy.

production and consumption levels, system stability, connectivity with other systems, and the degree of self-sufficiency; and
- Sustainable development, i.e., acceptance of the concept at the state level, the ratification of international agreements, the parameters of the national strategy of sustainable development, and the list of priorities.

Accordingly, *sustainable energy management can be defined as the process of energy management that is based on fundamental principles of sustainable development.* Sustainable management of energy must therefore be regarded as a concept that greatly differs from traditional management, which has largely ignored the need for conservation, restoration, and saving of energy resources. Thus developed, the concept of sustainable energy management is a major global change whose effects are felt throughout the world, and major changes in behavior and thinking of responsible institutions and individuals are required. Therefore, the concept of sustainable energy management can be seen as a challenge of modern mankind, which opens up possibilities for proposals and development of numerous solutions that will have long-term consequences for the development of human society as a whole.

The only possible way to approach the problem of sustainable energy management is by respecting the scientific achievements and practical experience as well as taking into consideration the particular characteristics of the subject that initiates, accepts, and implements the energy management in a sustainable way. In general, sustainable energy management involves the implementation of a number of different activities aimed at setting concrete goals of sustainable development into practice of production, transmission, and consumption of energy.

From the point of view of traditionally measured social and economic development, most important is the implementation of those measures and activities that bring about the most effective results in less time, but with respect to the principles of sustainability the imperative of time becomes less significant. Activities that bring quality results are acceptable, regardless of the time needed to detect and measure the positive effects.

Given the complexity and global scope of modern business, and taking into account a number of problems and disagreements in terms of global trends and priorities of development, sustainable energy management cannot be defined through a series of strictly determined activities that need to be undertaken. Modern science and contemporary practice show that the activities in the field of strategic and operational management, as well as the improvement of quality and socially responsible business, have become imperative in terms of implementation of the concept of sustainable energy management.

Theoretically, the problem represents the "gap between current and desired state." With this in mind, it is necessary to implement appropriate changes and undertake appropriate activities that would lead a company or community to an energy sustainable operation and therefore resolve the problems of that kind. The stages of more responsible operations are the following (Martins, Silva, Pimentel, Casias, & Campilho, 2020):

- Identification of problems (as a diagnostic activity) is a specific stage for each company because each company is characterized by a number of differences and particularities in relation to other companies. If the management of the company sets as the desired state the introduction of energy-sustainable business to a certain level (or completely), it should be in line with the ability to estimate its own strengths and define its own weaknesses. This is the stage of setting goals in the company that largely determines all other activities.
- Identification of development options includes prediction of activities that can be implemented to achieve the goals that relate to energy-efficient business operations. In predicting the ways in which it is possible to achieve the set goals, each option should be carefully elaborated and realistically estimated. At this stage, the experience of other companies in a similar situation can be useful (but not crucial).
- The selection of the most appropriate option leads to achieving the goal—energy-responsible business. This stage involves making appropriate decisions (among several options offered) that determine the future course of action.
- Implementation of the chosen method to achieve the goal includes a series of activities and requires the engagement of certain human resources, financial resources, and time.
- Control and correction of the identified deficiencies are a necessary activity that must be continuously carried out to correct irregularities and weaknesses that appear no matter how detailed the activities were planned.
- Achieving the goal, by which the company achieves a higher level of energy responsibility, about which all interested parties should be informed, mainly consumers, owners, and the community.
- Identification of (new) problems, thus returning the activities to the beginning, and the company strives to ensure that the activities on the improvement of ecologically suitable business are continuous.

In each company, the relevant service needs to monitor the situation in the environment, which is primarily related to new requirements for energy-responsible business (mostly coming from consumers and legislators) and to initiate the previously mentioned activities. In all stages leading to the achievement of environmentally responsible businesses, the whole management team should be involved, and if necessary, professionals can be engaged (Mickovic & Wouters, 2020).

Sustainable development itself is a kind of strategy of development of mankind that is not limited by time. It is subject to changes and adjustments. The effects of such development can be precisely measured at a certain point, but unlike traditional management, there is no strict orientation toward reaching goals. In fact, sustainable development is not defined in time and space and involves constant improvements so that it is not possible to reach a certain goal and then stop the implementation of the sustainable development management process. Sustainable development is not a goal in itself. It is a process whose goals are constantly redefined and corrected.

Thus, defining the process of sustainable development imposes the need to apply certain traditional strategic management activities, which enable implementation and control of the process in an unlimited period of time. Sustainable energy management is not a simple application of traditional strategic management, but implies its modification, which has led to the development of a unique model of sustainable management that differs from the traditional one. At the same time, it significantly questions the necessity of its implementation in the case of sustainable management in the field of energy (Trianni, Cagno, Bertolotti, Thollander, & Andersson, 2019).

To design a strategy for sustainable energy development, it is necessary to set specific priorities for development. Five basic priority programs have been suggested, which are diverse in program content but complementary from the point of coordinating the operation and development of the whole energy system, i.e., energy production sectors and the sectors of energy consumption and a gradual but consistent achievement of the goals promoted during the implementation of the suggested strategy. The priorities are as follows:

- The first priority of continuity of the technological modernization of existing power plants/systems/sources, in the sectors of oil, natural gas, coal from surface and underground mining, electric power sector, with production plants and facilities: thermal, hydro, and heating power plants and transmission system or distribution systems, and a sector of heating energy—city heating plants and industrial power plants.
- The second priority of rational use of quality energy and the increase of energy efficiency in production, distribution, and use of energy by the end users of energy services.

- The third-specific priority of using NRSE (new renewable energy sources) and new energy-efficient and environmentally friendly energy technologies and devices/equipment for energy use.
- The fourth-optional priority for the emergency/urgent investment in new power sources, with new gas technologies (combined gas-steam thermal power plants).
- The fifth long-term development and regional strategic priority, building new energy infrastructure facilities and electric power and heat sources within the energy sector, as well as capital-intensive energy infrastructure.

In accordance with the stated priorities whose implementation enables achievement of sustainable energy systems in the world, it is necessary to undertake the implementation of local and national programs that will be defined in accordance with the new approach to strategic management of energy. However, the traditional approach to strategic management is not fully acceptable for the purposes of sustainable management.

The actual results of the planning process should be concrete actions (implementation of the strategy). Planning decisions are only an intermediate stage of the planning process. Through the implementation of the strategy, the company transfers ideas into concrete achievements. In the process of strategy development, entrepreneurship and visionary qualities of leaders are required, while in the process of implementation, human resource skills are needed to achieve the set goals. Modern business is characterized by continuous change. The scope of the problem caused by change and the time required to solve the problem determine the complexity and the speed of implementation of the strategy. Depending on the size of the company and the style of management, there are several approaches to implementing the strategy. Each of the approaches has advantages and disadvantages, and the task of strategic management is depending on the scope of changes caused by the implementation of the selected strategy and other conditions, to decide the way in which the strategy will be implemented (Ma et al., 2018).

The implementation of strategic orientation is not a single process, but a process that involves the provision and allocation of resources, creation of organizational, procedural, motivational, and other conditions, and development of a series of individual plans for undertaking and coordinating activities for the achievement of these goals. The implementation of strategy involves a complex process of creating conditions and integrating activities to achieve results. For the proper management of activities, it is necessary to specify the responsibility and authority to identify tasks and budget and allocate them to the carriers through a system of coordination and overall management of the implementation process. In this case, emphasis is placed on global issues affecting the implementation of the strategy.

Sustainable energy management is an extremely complex process, both in planning and implementation (Jovanović, Filipović, & Bakić, 2017). The very implementation of sustainable energy management in practice has the following main characteristics that define it and distinguish it from traditional management:

- The implementation is extremely demanding. The transition from traditional to sustainable energy management requires the adoption and implementation of a number of major changes, both in thinking and in the manner of implementation and measurement of financial effects. If all planned activities are performed in accordance with scientific and practical recommendations, there is still a high degree of probability that the realization itself will not be fully effective and will not achieve certain goals. Managerial practice implies that implementation that reaches at least two-thirds of the defined goals can be considered very successful. When it comes to such complex changes, such as the transition to sustainable energy management, the degree of success is usually even lower.
- The implementation is a long-term process. Every large strategic change in any sphere of human activity requires enough time and its effects are most often visible after more than 10 years. The transition to sustainable energy management in this regard is not an exception. It can even be said that according to the time criterion, such a strategy change is even more demanding. The planned activities usually last at least 10 years, and the implementation itself, after which you can measure the effects, lasts at least the same period of time. Therefore, sustainable energy management is a process that requires extremely careful management, commitment, accuracy, continuity, and a clear commitment of all participants.
- The implementation is multidisciplinary. The problem of providing enough energy for the needs of modern mankind is one of the largest problems today. It will largely determine the direction and intensity of development of the planet in the future. The very acceptance of the concept of sustainable development takes a lot of time and is linked with a number of difficulties, and the transition to sustainable energy management is just one of the goals of sustainable development, which is also linked with a number of political, economic, and other characteristics. It is necessary to involve a number of stakeholders both on a local

and a national level. The problem of adequate energy supply that is not too expensive, will not cause too much pollution, and will be sufficient in both quantity and quality is one of the most complex problems of modern development, because energy directly affects the possibility of realization of most human activities. Therefore, the solution to this problem must involve all stakeholders: industry, services, agriculture, legislation, executive authorities, and all participants in the process of solving economic, environmental, energy, development, social, and other issues.

- The implementation requires strong IT support. The traditional way of managing the work of the energy sector includes the mandatory application of modern IT tools and software support, but transition to sustainable energy management is an even more demanding process. It includes specific modeling of relationships between production and consumption of energy from different sources and monitoring of the pollution. The modern trend imposes the need for rational use of energy and energy conservation at all levels. The support provided by information technologies is the only tool of support.
- The implementation requires a specific engagement of human resources. So far energy management processes have not been particularly demanding in terms of engagement of human resources. The involvement of professionals in the energy sector is technically necessary and therefore enough. Essentially, sustainable development integrates primarily environmental, economic, and energy knowledge and requires the engagement of experts from these areas. In addition, it is necessary to engage people who are fully trained to monitor issues of sustainable development and, accordingly, modeling of the development of the energy sector.
- The implementation is subject to many influences. The development of mankind and activities of people have always involved the consumption of certain amounts of energy. Progress has never been possible without an adequate amount of energy available. On the other hand, the progress of mankind has led to a large number of economic, ecological, cultural, and many other problems. The energy crisis and a high dependence of man on the energy available emphasize the problems in this field and make them top priority. Future development should be able to provide enough energy without causing excessive pollution. Any change in the energy field affects all actors of social and economic life, so their claims and efforts are subject to special attention. Modern processes of creation

of the development policy include consultation with all stakeholders, which further aggravates the already difficult transition to sustainable energy management.

- The implementation records negative financial effect at first. The need to move toward sustainable energy management was created primarily as a result of increased awareness of environmental pollution caused by production and consumption of energy. Necessary adaptation involves transition to new energy sources, which are mostly at the stage of examination, or their exploitation is insufficient so that a clear conclusion is not possible. The transition to a new way of energy management is necessary and occurs at the moment when there are not necessary prerequisites. Some technical solutions are not effective enough, exploitation is uncertain, distribution is difficult, legislation is often inadequate, and there are no financial investments planned in advance that are extremely high. Sustainable energy management was created as a result of the consciousness of mankind that the energy is a limited resource; the shift has already started, although there are no clear solutions to various problems encountered in the transition process. In many cases, the legislation imposes the transition to the production of certain amounts of energy from renewable sources without providing ways to realize that.

Therefore, the transition to sustainable energy management imposes the need for substantial financial investments, where there are no clear predictions of how long it will take for these investments to return the investment, although it is quite clear that all investments of this type are long-term. Therefore, the transition to sustainable energy management in the early years usually records more or less negative financial effects, because it requires significant investment, and production of energy from these sources is small and therefore generates low revenues (Polzin & Sanders, 2020).

Strategic analysis of energy sector

Strategic analysis is the initial phase in the field of energy management, both in the traditional and proposed model of sustainable management of energy. Various models are analyzed in various ways. Thus, analyses vary in the levels of detail, comprehensiveness, and the use of different temporal, geographical, political, and many other restrictions. Regardless of the scope and limitations, strategic analysis must be given special attention, because on the basis of this analysis, further steps and management procedures are taken, and any error in the process of analysis significantly affects the final result of the management process, most often after a period of many years (Boblenz, Frank, & Meyer, 2019).

For the purpose of a comprehensive and realistic understanding of the situation and subsequent creation of the appropriate strategy of the sustainable energy development, three basic analyses are necessary:

Global analysis—analysis of energy production at the global level is the first necessary condition for successful planning and implementation of the process of sustainable energy management. Energy is produced all around the world from different sources, in different quantities, and in different ways. Until the first half of the 20th century, energy was mostly spent in the place of production, i.e., it was transferred to shorter distances. With the increase in the number of people, globalization of the world economy, and exploitation of resources that are concentrated in certain parts of the world, the need emerged to produce increasing amounts of energy. Energy is becoming one of the key limiting factors of life on the planet, and it is needed to perform daily activities and complex human activities. Energy needs are constantly increasing, while the production of energy is trying to meet ever-increasing demands. Because of this, energy has become the subject of world trade and an important factor in business and political changes. In addition, awareness of the problem of limited energy resources raises the issue of energy production to a higher level, and energy management is becoming one of the key issues of sustainable development and development in general.

Global analysis includes the following (Duan & Chen, 2018):

- Analysis of global consumption—consumption of energy resources in total, consumption in the regions and in the countries, consumption of energy according to energy types, consumption according to the purpose, average energy efficiency, energy efficiency according to the fields of industry, economic parameters of energy consumption; and
- Analysis of global production—global reserves of energy resources according to the types and distribution, availability of the renewable energy resources, production of energy from renewable resources, global trends, economic parameters of the world energy industry.

National analysis—apart from the analysis of the global energy situation, availability of resources, production, and consumption, it is particularly important to make adequate analysis of the situation in the field of energy on the national level.

National analysis should include the following (Fattahi, Sijm, & Faaij, 2020):

- Analysis of national consumption—quantity of energy spent in a certain country in a certain period of time, in total, according to the types of energy and field of activity, definition of the biggest consumers of energy in the regions and according to the field of activity, energy efficiency according to the field of activity;
- Analysis of national production—available renewable and nonrenewable energy resources, legislation on exploitation, structure of ownership and capital, existing technologies, impact on the environment, regulation of energy waste;
- The degree of self-sufficiency—a real estimation of self-sufficiency is particularly important in creation of national strategy of energy development. Countries that can satisfy their needs for energy are few, and their monopoly provides them the opportunity to influence world politics in all domains. Most countries in the world are forced to import energy of different types and in different quantities and because of this are forced to adjust themselves to different influences; and
- Harmful emissions into the environment—consumption of energy resources necessarily leads to a variety of emissions into the environment. These emissions are the focus of observation of all stakeholders because of the long-term adverse impacts on the greenhouse effect and global warming. The analysis of the ecological cost of progress should be made, i.e., it is necessary to determine the load of the environment per GDP.

Analysis of the company's current—a global strategy of sustainable development and global availability of energy resources certainly influences the framework in which a certain company operates or will operate, since the issue of energy industry is long-term and complex, but the situation inside the company is crucial to solving the problem, because companies are just agents who, along with individual consumption, largely affect the energy future of the world (Gu, Wei, Zhang, & Yan, 2019).

From the point of view of the company, the analysis should include:

- Internal analysis—energy consumption by type and quantity, in total, per unit, per unit of revenue, the advantages and disadvantages of using certain energy, technological solutions, possible improvements, harmful emission into the environment; and
- External analysis—analysis of the energy situation in the immediate environment of the company, especially energy policy of suppliers, subcontractors, competitors, attitude and activities of the local community, support of financial institutions, tax incentives, special credit lines,

subsidies, and the possibility of obtaining the necessary energy from renewable sources.

Global analysis offers general guidelines for creating environmental policies in each country. On the basis of general guidelines and specific analysis on a national level, each country develops its own strategy of sustainable development and, within it, a special strategy of energy development at the national level. Proper and realistic analysis is the basic precondition for the creation of adequate energy strategy, which recognizes the concept of sustainable development and will be the basis for all long-term activities in the sphere of energy production and consumption. Only on the basis of correct analysis of the situation is it possible to define the desired goals and activities needed to achieve these goals.

The analysis of the current situation in the company is of particular importance. The energy policy of any responsible company should be harmonized with the national strategy of sustainable development and the strategy of energy development, its basic principles, and priorities. Each company has to conduct its own analysis and come to certain conclusions as well as establish its particular objectives of sustainable energy development (Méndez-Piñero & Colón-Vázquez, 2013). Apart from this, only with a realistic assessment of the situation can a company define the resources required to achieve the defined goals with great precision (equipment, money, people, and time).

Strategic objectives of sustainable energy management

Energy management has always been a challenge for anyone involved in the design and implementation of the process, and in most cases, the consequences of energy management are far-reaching. The complexity of the global economic and environmental situation and the full awareness that there is not enough energy, as well as that the processes of generation, transmission, and consumption of energy, which are the greatest polluters of the environment, have caused minor and major ecological disasters and led to major changes in traditional energy management.

The goals of traditional energy management were based primarily on the principles of economic growth and the demands of society, and the main objectives of the owners of energy resources were to find effective ways for successful and cheaper exploitation of energy resources. The next goal related to providing storage of certain resources until delivery, ensuring adequate safety standards, which were not related to the protection of the environment, but were primarily aimed at preserving stocks of resources. In addition, an important goal of the owners of energy resources was reflected in the fact that resources were considered the owner's property (company, state, private capital) and were sold as commodities in the world market at the highest possible price. Exploitation in order to gain economic profit and monopoly position was the only business objective of the owner of the energy resources (Becker, Demski, Evensen, & Pidgeon, 2019).

Similarly, there were certain goals regarding energy transfer. Traditionally, the main goal of this segment of the energy management process was to transport the energy resource, no matter whether it was in a processed, semi-processed, or raw form, in a way that was most economical. Fast and efficient transport of energy in order to gain a competitive advantage was the only objective that guided the energy transfer processes.

Energy distribution in the traditional sense means the supply of a certain type of energy to end users. Accordingly, the main objective of traditionally based distribution of energy lied in the fact that the energy was bought at an adequate price and stored, transported, and sold to customers at a price satisfactory to the energy carrier.

The end users, as users of energy, set as the main objective the ability to buy certain energy resources at an adequate price and to make the supply stable. Production processes, as well as the products and services, are designed to use energy in a way that is conditioned by technology, without considering the possible savings. The value of energy that is consumed in the production process is compensated by the final price of products or services that are paid by the consumers.

As with the owners of energy resources, economic interest was the main goal of business. In accordance with that, all steps of management were defined, decisions were made, technologies were installed, and the whole system was defined by certain legislation that in a traditional way regulated the relationship among all participants in the process of energy supply. Thus, the defined system of energy production and consumption survived proportionally long, although with the start of the first energy crisis (in the 1970s), all shortcomings of such energy management started to be identified.

The emergence of environmental problems, warnings about the limited supply of energy resources, pollution caused by the use of energy resources, and the increase in consumer awareness have led to the need to introduce significant changes in the process of treatment of energy from producers to end users. Modern sustainable energy management has emerged as a unique phenomenon with its own set of goals.

The main objective of sustainable energy management lies in the fact that energy is produced and consumed in ways that will not endanger the opportunities of future generations to meet their energy needs. Since economic growth is closely related to sufficient quantities of available energy, the main goal of energy management can be defined as energy management that will allow economic growth that will not burden the environment over the limits of sustainability (Mangla et al., 2020).

Other objectives derived from this objective are the following:

- To present the relationship between ecology, economy, and energy as a very important element of economic and political stability;
- To promote a public information system regarding the need to preserve the environment through sustainable management of energy resources (including renewable energy—small hydro plants, wind generators, geothermal sources, solar energy, and biogas), with a strong emphasis on the policy of national development and regional cooperation;
- To raise consumer awareness about the fact that energy use has a major impact on environmental issues and economic development;
- To encourage dialogue between various stakeholders in the field of energy and ecology, with special emphasis on rural areas where the production of energy from renewable sources can be used as an opportunity for job creation;
- To participate in creating a national energy strategy that will accept the contemporary experience of other countries in the region and European energy policy;
- To support the development of small and medium enterprises (SMEs), particularly in the production of energy from renewable sources;
- To promote sustainable development by appropriate policy and the use of renewable energy sources;
- To assist local initiatives in the process of public participation in decision-making, support the implementation of appropriate procedures, and provide access to information as well as implementation of legislation;
- To establish strong communication with relevant government and other institutions in order to participate in the development of energy strategy. Joint working groups or committees should be established at the operational level to facilitate intersectorial cooperation;
- To provide access to information and education in the field of renewable energy sources to all stakeholders (different levels of management and decision-makers, professional institutions, NGOs, interest groups, and

organizations) through the establishment of educational centers for renewable energy sources;

- To develop and implement projects at the local level in the function of education of all stakeholders in the process of planning and decision-making and participation of the local business sector and local communities, in the broadest way;
- To initiate the establishment of a national network for management of renewable energy sources that will work on promoting the use of renewable resources and programs for conservation and efficient use of energy. The network should also support access to finance for small and medium-sized producers of energy;
- To mobilize financial institutions to encourage investments in this sector;
- To work on promoting measures that will provide support through fiscal and financial policy (lower taxes, lower interest rates); and
- To support research and development of energy efficient products and technologies.

Case study

From tip to toes: Mapping community energy models in Canada and New Zealand. Energy Policy, 121: 162–174. https://doi.org/10.1016/j.enpol.2018.06.002.

References

Adizes, I. (2018). *Insights on management: Volume III.* Adizes Institute Publications.
Andrews, K. (1987). *The concept of corporate strategy.* Richard D Irwin; Subsequent edition.
Ansoff, I. (1979). *Strategic management.* Palgrave Macmillan.
Becker, S., Demski, C., Evensen, D., & Pidgeon, N. (2019). Of profits, transparency, and responsibility: Public views on financing energy system change in Great Britain. *Energy Research & Social Science, 55*, 236–246.
Boblenz, K., Frank, V., & Meyer, B. (2019). Energy system analysis for evaluation of sector coupling technologies. *Fuel, 254*, 115658.
Capehart, B. L., Turner, W. C., & Kennedy, W. J. (2016). *Guide to energy management* (8th ed.). Routledge, Taylor & Francis Group.
Certo, S. C., & Peter, J. P. (1989). *Strategic management.* New York: McGraw-Hill Inc.
David, F. R., David, F. R., & David, M. E. (2020). *Strategic management: A competitive advantage approach, concepts and cases* (17th ed.). Pearson.
Duan, C., & Chen, B. (2018). Analysis of global energy consumption inequality by using Lorenz curve. *Energy Procedia, 152*, 750–755.
Elgersma, E. (2017). *The strategic analysis cycle hand book: How advanced data collection and analysis underpins winning strategies.* LID Publishing.

Fattahi, A., Sijm, J., & Faaij, A. (2020). A systemic approach to analyze integrated energy system modeling tools: A review of national models. *Renewable and Sustainable Energy Reviews, 133*, 110195.
Gu, W., Wei, L., Zhang, W., & Yan, X. (2019). Evolutionary game analysis of cooperation between natural resource- and energy-intensive companies in reverse logistics operations. *International Journal of Production Economics, 218*, 159–169.
Johnson, G., Scholes, K., & Whittington, R. (2008). *Exploring corporate strategy* (8th ed.). Pearson College Div.
Jovanović, B., Filipović, J., & Bakić, V. (2017). Energy management system implementation in Serbian manufacturing—Plan-do-check-act cycle approach. *Journal of Cleaner Production, 162*, 1144–1156.
Ma, T., Wu, J., Hao, L., Lee, W. J., Yan, H., & Li, D. (2018). The optimal structure planning and energy management strategies of smart multi energy systems. *Energy, 160*, 122–141.
Mangla, S. K., Luthra, S., Jakhar, S., Gandhi, S., Muduli, K., & Kumar, A. (2020). A step to clean energy—Sustainability in energy system management in an emerging economy context. *Journal of Cleaner Production, 242*, 118462.
Martins, L., Silva, F. J. G., Pimentel, C., Casias, R. B., & Campilho, R. D. S. G. (2020). Improving preventive maintenance management in an energy solutions company. *Procedia Manufacturing, 51*, 1551–1558.
Méndez-Piñero, M. I., & Colón-Vázquez, M. (2013). Economic analysis of alternatives for optimizing energy use in manufacturing companies. *Energy Economics, 40*, 146–154.
Mickovic, A., & Wouters, M. (2020). Energy costs information in manufacturing companies: A systematic literature review. *Journal of Cleaner Production, 254*, 119927.
Ohmae, K. (1982). *The mind of the strategist: The art of Japanese business*. McGraw-Hill Education.
Polzin, F., & Sanders, M. (2020). How to finance the transition to low-carbon energy in Europe? *Energy Policy, 147*, 111863.
Radovanović (Golušin), M., Dodić, S., & Popov, S. (2013). *Sustainable energy management* (1st ed.). Elsevier—Oxford Academic Press.
Robbins, S. P., Judge, T. A., & Vohra, N. (2016). *Organization behavior* (16th ed.). Pearson Education.
Rossa, S., Doty, S., & Turner, W. (2018). *Energy management handbook* (9th ed.). River Publishers.
Sapmpler, J. L. (2021). *Bringing strategy back: How strategic shock absorbers make planning relevant in a world of constant change*. Gildan Audio and Blackstone Publishing. Unabridged edition.
Smith, C. B., & Parmenter, K. E. (2015). *Energy management principles: Applications, benefits, savings* (2nd ed.). Elsevier.
Stoner, J. A. F., Freeman, R. E., Gilbert, D. R., & Nambudiri, R. (2018). *Management* (6th ed.). India: Pearson Education.
Trianni, A., Cagno, E., Bertolotti, M., Thollander, P., & Andersson, E. (2019). Energy management: A practice-based assessment model. *Applied Energy, 235*, 1614–1636.
Vincze, J. W., & Higgins, J. M. (1992). *Strategic management: Text and cases*. Harcourt School.

CHAPTER 5

Implementation of sustainable energy management

Contents

Development of strategy of sustainable management of energy

The implementation of strategy is to the core step in the process of strategic management. No matter how difficult it is to develop the strategy for sustainable energy management, it is even more difficult to implement it. The strategy is the planning framework for taking actions that would use

Sustainable Energy Management
https://doi.org/10.1016/B978-0-12-821086-4.00003-9

energy resources adequately. Therefore, the implementation of the sustainable management of energy is an important field in strategic management (Javied & Franke, 2015). The strategy is first developed and then executed, but the danger for the (lack of) success exists in both areas, in the process of developing a strategy and the process of implementing the strategy, as is illustrated in Fig. 5.1.

In developing strategy, first it is necessary to consider the scope of the problem and the time available for its solution. The transition to sustainable management of energy is a significant change, which depends on the intensity of energy consumption in the company, but also on the method of management and current characteristic of organizational culture in the indicated company (König, 2020). The strategy can be planned and implemented in different ways, with different final outcomes:

- Success occurs when the strategy of sustainable energy management is developed well and implemented well, which means the internal and external opportunities have been identified on time and organizational, resource, motivational, and other conditions have been created for the implementation of the strategy. This situation is certainly the most desirable for the successful process of sustainable energy management.
- A roulette situation is characterized by a poorly developed but well-implemented strategy of sustainable energy management, where deficiencies have been identified during the implementation of the strategy and can be surpassed. This situation can be found relatively often in practice and shows that there are still major problems in the planning and implementation of sustainable energy management. This situation can be partly justified by the fact that the transition to sustainable energy management is a huge undertaking and includes a great number of uncertainties.

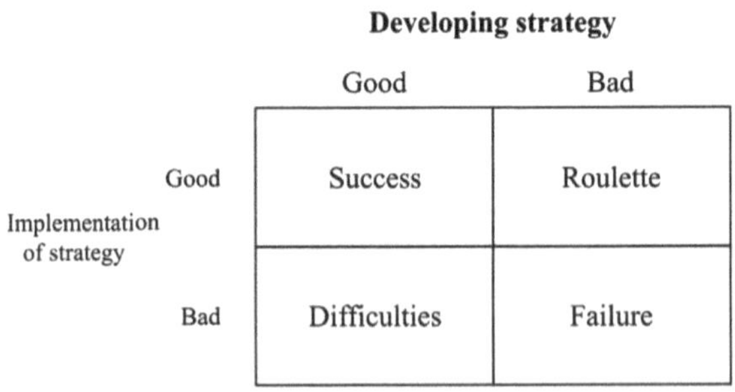

Fig. 5.1 The relationship between the implementation and development of strategy.

- Difficulties arise if the strategy is well formulated but poorly implemented. It often happens in practice, but in this case the organization would change the existing strategy rather than remove barriers that prevent implementation of the strategy. Taking into account the fact that the strategy of sustainable energy management is a very long process, this situation is particularly undesirable.
- Failure is a situation that occurs when a poorly formulated strategy is poorly implemented, leading to confusion and failure. This case can hardly be corrected. In an attempt to create and implement strategies for sustainable energy management, this is certainly the least desirable situation and involves a great number of financial, environmental, organizational, and other problems whose impact is great and long-term.

The actual result of the process of planning sustainable energy management should be concrete actions (implementation strategies), while planned decisions are only a by-product of planning. Through the implementation of the strategy, the company transforms ideas into concrete achievements. This process requires that the goals are transformed into concrete tasks, activities are identified, the necessary resources are allocated, and the business culture is integrated.

In the process of developing a strategy, entrepreneurship and visionary leadership qualities are emphasized, and in the process of implementation, it is important to use human resources to achieve the goals. Therefore, the implementation of energy strategy means taking actions to bring into practice the intentions stipulated by the mission and goals. The implementation runs the activities of the company, i.e., people, technology, and resources, and the management of the company has to anticipate possible problems, conflicts, production bottlenecks, and other external and internal barriers.

Modern business is characterized by continuous change. Changes are quantitatively and qualitatively significant and far-reaching in the field of energy production and consumption. These changes can be different: a change in the size of company, mergers and acquisitions, new competitors entering the industry, the emergence of substitutes, changes in the organizational structure of enterprises, changing market conditions, and a number of other factors (Casco-Vidrean, Niemann, Baidoc, & Pisla, 2018). The existence of several large global companies for the exploitation of energy resources, powerful monopolies, and the growing demand additionally influence the process of creating a sustainable energy management. It can be said that the 21st century is a century of change, and that change is

		Time needed for the change of implementation: Long	Time needed for the change of implementation: Short
The scope of strategy problem	Big	Sequential interventions	Complete interventions
	Small	Evolutionary interventions	Managerial interventions

Fig. 5.2 Types of strategy implementations.

the only certainty. The scope of the problem and the urgency to solve the problems influence the complexity and the speed of implementation of strategies, as shown in Fig. 5.2.

When a company is faced with small problems and the period of time available is long enough, the implementation of the energy strategy would require a shorter period of time. At the same time, the implementation of strategy will include evolutionary or gradual (incremental) changes. This situation is characteristic for enterprises and sectors that are not highly energy-dependent. In such cases, the company has enough time to design, plan, and carry out certain activities in the field of energy management (Ahmad, Jan, Jabeen, & Alvarado, 2021). These companies are potentially in the best position, because they have enough time to understand the specifics of the transition from traditional to sustainable energy management, to quantify the necessary changes, and to predict results and timely direct the company in the right direction. They have time enough, given the low energy demand, to foresee and implement the use of renewable energy sources, to develop activities to improve energy efficiency, and to reduce the energy waste to a minimum. In other words, these companies can and should use all the available interventions and evolutionary mechanisms for systematic resolution of the problem of energy solving. In that way, companies are able to gradually transfer to a more acceptable way of using energy, without endangering their own stability.

Another situation is typical for companies that consume less energy but for some reason are forced to react quickly. If the problems the company faces are small, but should be resolved in a short period of time, the activities are performed by management (management intervention) directly at places where the problems emerged. As modern business is characterized by rapid changes that often result in serious problems, companies are often in a situation to respond quickly to certain influences. Factors that often force a company to launch its managerial mechanisms and adjust its own energy strategy in a relatively short period of time include sudden changes in energy prices, changes in legislation, the pressure of the local community, filed lawsuits, environmental pollution, and the like (Kuo & Smith, 2018). The response of management must be quick, but certainly in conformity with the long-term global orientation of companies on sustainability in the field of energy needs. Of particular importance is that modern companies, regardless of their size or energy intensity, integrate their strategy of energy management into the existing strategic plan for the development of the company and that any decision concerning changes in the business is compliant with the adopted view that sustainable energy is the only way of energy management in the future. Not a single goal of the company must be set, if, among other things, it does not comply with this requirement.

When there are major problems, the company must react in a different way. This situation usually happens in companies that consume large amounts of energy. Depending on how much time the organization has to solve the problem, there are two strategies. If the company has a long period of time at its disposal and the problems are huge, it is necessary to react sequentially, i.e., in stages. In fact, most modern companies are facing a gradual need to switch to sustainable energy management and thus have at their disposal a certain period of time to adjust its business operations to it. Most countries have legislation that stipulates quite a long period of time to solve this problem. In this case, it is optimal that the enterprise acts on the principle of sequential (gradual, phase) interventions because in that way they can optimize the use of its resources and gradually resolve the existing problems (Liu, Fang, & Xie, 2021). Usually, the company first adjusts its own rules and regulations on business operations, which should be adjusted to positive legislation. Then, the company defines its own energy strategy within which it determines its own vision and mission and adjusts its goals to the principle of sustainability. After that, the company gradually moves on to solving concrete problems (introduction of energy consumption from renewable sources, improving energy efficiency, implementation of energy savings, etc.), depending on its capabilities, needs, and support provided by the environment.

However, if there are problems that cannot wait, the company has to react quickly. Most commonly, this situation happens in companies that are large consumers of energy and suddenly find themselves in a situation in which a change in the approach to energy management is required. The reason is usually pressure from the environment, where the changes of energy prices and lawsuits are the driving factors for the company to react urgently. In these cases, adequate complete intervention is applied, while the management of the company must fit (synchronize) changes in all parts of the company. Solving of the problem in this way is connected with increased risk and often represents a hazardous undertaking, requiring considerable experience of decision-makers (Xia, Wei, Gao, & Ma, 2020). The sudden switch to another type of energy, the termination of cooperation with existing suppliers, and the attempt to avoid paying fines or similar in most cases cannot be resolved successfully in this way. The only way to avoid this approach in problem-solving is to avoid having to solve huge problems in the field of energy in a short period of time. This is possible only if the strategy is directed in a timely manner according to the principle of sustainability.

The following ways of reacting to problems can be practically realized only by introducing changes in management of the company. Depending on the size of the company and management style, there are several approaches to implementing the strategy of sustainable energy management, as follows:

- The command approach is typical of smaller companies or of situations in which the implementation of the strategy does not involve a lot of changes. This approach is based on the fact that the formulated strategy is executed by lower levels of management. In fact, the tasks of strategic management should be separated into "creative thinking" and "implementing" operations. These cases are usually detected in companies that are small consumers of energy. Creating a plan of transition to sustainable energy management is simple, and realization of goals does not require the increased use of the resources of the company. A company can fully convert to the use of sustainable energy sources without specific changes that would threaten the stability of the business.
- The organizational change approach requires greater activity of the top management, thorough understanding, and designing of the necessary changes in organization, planning, rewards, and control. This approach takes into account the need for forecasting and applying various organizational and other changes in implementation of the strategy. The above case is specific to companies that consume large amounts of energy and are aware of the need to change the energy strategy, but for some reason, such determination is not known to all employees. In these cases, company management makes decisions and foresees necessary changes that are further implemented without any special effort to familiarize all employees with them. This approach is particularly effective when companies are in a position to have to quickly change their energy

strategy due to particular pressure, while management has neither the time nor the mechanisms to inform all employees.

- The collaboration approach is a process where managers of business functions and units are involved in the process of implementation so as to obtain more realistic estimates of implementation tasks. This approach motivates managers and encourages them to take on greater responsibility for the implementation. This approach is possible in cases when companies are small or large consumers of energy, but what they have in common is that they have enough time to switch to sustainable energy management. This approach is, however, exclusively reserved for companies with strong management structure, where there is no comprehensive inclusion of all employees in planning and implementation of change.
- The cultural approach is a process of designing a strategy and its implementation that includes all those to which the strategy applies. This is a broader approach than the previous one and is based on good communication and encouraging employees to identify their culture with the mission and vision of the organization, which will result in better operationalization of the strategy. This approach is suitable for organizations that insist on the business culture and have rapid growth. Usually, these are organizations that carry out activities in sectors with high technology. This approach is also an excellent solution in organizations whose long-term development strategies are aimed toward corporate social responsibility, where there is an appropriate organizational culture and practice of involvement of all employees in all aspects of business. The transition to sustainable energy management essentially considers these approaches an ideal solution.
- The bottom-up approach insists on more intensive engagement of the direct executives in designing and implementation of the strategy. In this way, the executives are encouraged to have creative thinking and initiative. Top management has the role of auditors of the proposed strategy of sustainable energy management in the company.

Obviously, each of these approaches has advantages and disadvantages, and the task of management is to create a flow of implementing the strategy, depending on the size of the changes caused by the implementation. For the implementation of any approach, it is very important to clarify the solutions through communication, to operationalize the objectives, and to encourage further development and motivation of the executives. The transition from a traditional to sustainable approach to energy management is a complex and long-term process that greatly influences the stability of business operations and therefore requires special attention of all persons involved in the decision-making process (Noboa & Upham, 2018).

Traditional approaches to implementation of sustainable energy management

Implementation of the sustainable management of energy refers to running activities and resources of the company so as to achieve the set goals in accordance with the strategic orientation. The transition from traditional to sustainable energy management is a major strategic change and involves huge utilization of all resources. Implementation of strategic orientation is not a single process, but a process that involves provision and allocation of resources, creation of organizational, procedural, motivational, and other conditions, and making a series of individual plans for undertaking and coordinating activities for the realization of the objectives, taking into account the sustainability of the environment (Chen, Xiong, Li, Sun, & Yang, 2019).

The implementation of the strategy of sustainable energy management is a never-ending process since the external and internal environment is continuously changing and completely new and unique situations occur that require understanding the relationship between the functions of the company. It is necessary to establish priorities, activities, and holders. All predicted activities should be properly connected and synchronized.

Creation of organizational structure for implementation of the strategy

This is an essential element for the implementation of the strategy that is used to create a new or adapt the existing organizational structure, which is required for the smooth implementation of the strategic directions and ways of doing business in the organization. At this stage, the focus is on establishing adequate internal organization and management skills that match the strategic determination and vision. The transition to and implementation of sustainable energy management involve major changes in the entire organization and acceptance of change by all employees (Si, Wang, & Kong, 2020). There are a number of strategic decisions that should be made at various levels of management, and these decisions need to be communicated to employees. Sustainable management of energy requires the full commitment of management, but employees should be familiar with the changes in order to change their way of thinking and take an active part in the new strategy.

Allocation and focusing of resources toward strategic goals

Since operationalization of the strategy requires engagement and using a number of resources, it is important to take into account the anticipated changes, the scope of activities, and provide a budget for each group of tasks and their appropriate use. It is necessary to understand the nature, importance, and priorities of different parts of the strategy and methodology of calculating costs and expenses. Determining the required resources and their allocation is the most crucial prerequisite for the implementation of the strategy of sustainable energy management. A particular problem arises in companies that are major producers and/or large consumers of energy. Their traditional way of work involves the use of energy without limits, in accordance with needs, and the payment of energy is through the price of the final product. Studies have shown that in the cost of a product, the value of energy is often unrealistically high. In general, poor indicators of energy efficiency give much room for savings (Frate et al., 2018). Besides the rationalization of energy consumption, companies are often faced with the objective costs of the transition to using a certain amount of energy from renewable sources, which requires investments that are not paid back quickly or easily. For these reasons, companies transferring to sustainable energy management must provide certain financial resources for these needs, from their own funds or by using other financial sources, including the use of financial sources and subsidies predicted in most countries, which adopted new legislations and legal mechanisms for promotion of sustainable energy development.

Galvanization of implementation of the chosen strategy

This stage includes understanding, maintaining, and coordinating overall activities of the company in order to implement the strategy and create the infrastructure needed to ensure efficient and effective implementation. The commitment of management and all employees is crucial to the success of implementation of the strategy for sustainable energy management. It is, therefore, necessary to design and develop specific mechanisms for monitoring and directing behavior of employees in the direction of the company goals. Most often, tools such as training programs, development of a system of rewards and punishments, and other mechanisms that are considered acceptable and effective in terms of employee motivation are used (Higuchi, Mhede, & Sonobe, 2019).

Establishment of an internal administrative system of support

This stage includes establishing the activities that institutionalize the implementation of the strategy by providing appropriate procedures, policies, and other internal regulations (Lima, McMahon, & Cabral Seixas Costa, 2020). Under the influence of new circumstances, new business policies are incorporated in revised procedures and efforts, including:

- Adapting policies and procedures for the support of the strategy, which clearly define the rules and procedures for the implementation of sustainable strategy. A clearly defined strategy of general sustainable development, as well as clearly defined procedures for carrying out the activities and delegated responsibilities, creates a great positive precondition for successful implementation of the strategy. Energy management in companies is a complex process and requires involvement of employees of different profiles, and clear definition of individual responsibilities and duties is of special importance.
- Ensuring an adequate information system, without which the transition to and implementation of sustainable energy management cannot be realized. Contemporary information systems have to be designed so they can accept all necessary input parameters, to process them, and to develop a specific portfolio of energy consumption that is assumed to be the most effective, i.e., that will be compatible with the objectives as much as possible. The information system must follow all the parameters of production and consumption by type and quantity, the places of consumption, and the amount of energy waste. Also it has to propose the ways of energy saving and to predict effects of energy saving. In addition, it must show all financial and other effects of the suggested changes.
- Institutionalizing the internal system of control implies creating and constantly improving the control system. Modern business imposes strengthening of the system of control, but it is an absolute imperative when it comes to the implementation of the strategy of sustainable energy management since it strongly affects all aspects and effects of business. Energy is a resource that considerably limits the business of a great number of companies, so the system of control should be properly structured and properly implemented. Strategic objectives must not remain objectives only. The success of implementation of a certain strategy should not be determined by the application of general measures and qualitative evaluations. Achieving goals and effects of the whole strategy of sustainable energy management can be monitored only if the goals are

measurable, and all parameters of control (the essence, indicators for monitoring, frequency) must be harmonized with the specific features of the existing and new system of control in the field of energy.

- Ensuring good relations between the strategy requirements and the current way of doing business. Development and implementation of the strategy of sustainable management of energy have to be integrated in all other development planes of the company. It should not be carried out suddenly, nor should the company set unrealistically high goals. Their implementation is usually impossible, bringing about great difficulties and the expenditure of time and resources, and failure often leads to the loss of motivation and temporary or permanent loss of vision for the company. The strategy of sustainable energy management has to be created in accordance with the current operations of the company, and all planned changes must be gradual and clearly oriented toward the goals.

All these basic managerial measures and mechanisms are the factors that define possibilities for successful implementation of the strategy of sustainable energy management in all its stages. The mistakes that often occur in this field include vaguely defined vision and mission, a lack of integration of the strategy of sustainable energy management into the existing strategic planes, unrealistic goals, and an inefficient system of control.

Providing strategic leadership

For the implementation of the sustainable management of energy, leadership is very important. Managers need to have entrepreneurial and leadership skills, because they are expected to inspire, create policies, resolve crises, allocate resources, represent the company well, etc. The leader is a person who, without being forced, has the ability to motivate and gather together employees based on the company's vision, ideas, and charisma, while managers, through planning, organizing, and control, ensure the running of the business (Graves, Sarkis, & Gold, 2019).

Implementation of the strategy of sustainable energy management is a complex process of creating conditions and integration of activities to achieve expected results. To manage these activities, it is necessary to specify the responsibility and authority and to identify the tasks and budget and allocate them to the carriers through a system of coordination and overall management of the implementation process. In this regard, special emphasis is placed on global issues affecting the implementation of the strategy, as follows:

- Creation of adequate organizational structure and corporate culture of the company;
- Transformation of long-term goals into the current levels of performances as well as allocation of tasks and budget to functional areas;
- Management of activities and management of the process of implementation; and
- Control and audit of strategy.

These factors represent only a few preconditions for successful implementation of the strategy of sustainable energy management, but theory and practice have shown that these factors have the strongest individual and synergy effects.

Creation of organizational structure and corporate culture

As the implementation of the strategy for sustainable management of energy involves mobilizing people and resources in a rational way, one of the key issues to be addressed is the establishment of an appropriate organizational infrastructure that can ensure harmonious operations of the company as a whole. If you understand the structure of an organization as a way of identifying the top and parts of the enterprise, as well as the relations between them, then its structure is a key element in the implementation of the strategy (Isensee, Teuteberg, Griese, & Topi, 2020). The implementation of the strategy of sustainable management of energy is by rule integrated into all organizational units of the company and as such creates additional organizational difficulties. Organizational structure coordinates the company philosophy and anatomy in an effort to ensure optimal working abilities and vitality in the specific economic conditions. Objectives, responsibilities, abilities, and tasks of employees should be coordinated, so as to create significant competitive advantage by implementing the strategy. In that way the system of support is provided for implementation of the activities in a value chain of the company. Only the optimal organizational structure can ensure the realization of strategic goals of energy management (Schneider, 2020). If employees are not adequately organized and motivated, implementation of the best-designed strategy of energy management should be questioned.

The organizational structure is not only an instrument for implementing the strategy, but also the basis for creating strategic performances in the company. The organizational structure is the infrastructure that provides a rational relation between objectives and those who perform the activities in the company. This is even more important when it comes to creating energy performances, which are long-term and often associated with high costs.

Today, businesses can be organizationally regulated in a relatively small number of ways. There are certain organizational structures, with minor or major modifications, that can be applied in business, regardless of the size of the company, the form of ownership, energy consumption, or the activity that the company deals with. Small businesses that are minor consumers of energy or large companies that consume huge amounts of energy can have the same organizational structure and be equally successful or unsuccessful in business. The impossibility of creating a large number of varieties of organizational structure is one of the limiting factors in trying to adapt the organizational structure to the new way of energy management. Building organizational structure is based on differentiation, coordination, and integration (Fadnavis, Najarzadeh, & Badurdeen, 2020).

Differentiation is a way of identifying parts of the company that will, according to similarity of businesses, provide opportunities to achieve the maximum effects of specialization and concentration. Differentiation of individual company units must be coordinated with all the functions of the company, not only with the needs of sustainable energy management. Additional problems arise from the stipulated fact regarding organizational differentiation that can be considered acceptable for sustainable management of energy. Very often the organizational structure of a company is adapted to the form of ownership or is determined by the activities the company does, but at the same time it can be completely inefficient for the needs of energy management.

Therefore, in most cases, at the beginning of the process of planning the strategy of sustainable management of energy, it is necessary to create a detailed review of the existing organizational structure and adjust it to new needs. The most effective technique for differentiation is often separation of parts of the company by activities or work processes performed in that part of the company. In order to consider some unit acceptable in terms of sustainable energy management, that unit needs to be described with certain energy parameters, independently of other parts of the enterprise.

If the selected organizational unit cannot be accurately described in this way, the stipulated unit may not be well differentiated. If the differentiated organizational unit, compared with other units or the company as a whole, can be described by certain indicators of energy consumption based on the quantity and/or the type of energy used, or based on energy efficiency or energy price that was spent in that unit, it can be considered that the stipulated unit is well differentiated and that being such, it can constitute a separate unit of the organizational structure (Brown, Pham, & Sivabalan, 2020).

Coordination involves establishing relationships between organizational units in order to provide synergy in business. Very often the differentiation of organizational units can be implemented successfully, but the organizational structure is not successful. In addition, it often happens that the same organizational structure in companies with the same scope of activities may show different effects. Therefore, it is clear that the best differentiated organizational structure cannot show its positive effects if the differentiated parts are not linked in an efficient way.

In theory and in practice, there are certain suggestions and solutions to connect the differentiated parts of the organizational structure. The success of the links between the elements of organizational structure therefore largely depends solely on the respect of these links (Addis, Capone, Carello, Gianoli, & Sansò, 2016). In certain operating systems, the organizational structure that is observed only in one way, as a bond that is directed from higher to lower levels, may prove to be most appropriate. However, planning and implementing the strategy for sustainable energy management require a strong one-way communication in the initial stage, but during the time, it is necessary to redefine the organizational structure and make it more flexible and two-way. Since each part of the organizational structure must be described with certain energy parameters, efficient connection between the parts is almost essential.

Integration refers to the detection of the number of hierarchical levels, their responsibilities and powers and grounded on that, provision of effective management and control. The companies that are large consumers of energy, and therefore particularly motivated to implement a strategy for sustainable management of energy, often have complex organizational structures and operations at more hierarchical levels (Farham, Mohammadian, Alipour, & Pouladi, 2019). Very often these are the companies that operate in many locations in several states or even on different continents. All this affects the need to properly and timely determine the levels of decision-making.

The strategy of sustainable energy management is primarily in the domain of activities of top management of companies, or the highest hierarchical level, but the practical realization and particularly the control are in the domain of lower hierarchical levels, and activities are conducted in the immediate realization of activities, at the lowest level that usually is not able to make decisions. The long-term impact, sensitivity, and economic value of changes in the energy sector give additional importance to the need to precisely determine the hierarchical levels of decision-making.

As the implementation of the strategy often leads to changes in size, structure, and activities of the company, and its bonds with the environment, it is necessary to anticipate the new organizational structure in the business. Internalization of business and geographical spread of activities change the existing relations between the top and the parts, so it is necessary to establish the centers of responsibility with new competencies. Implementation of the new strategy requires adjustment of the organizational structure; otherwise, it leads to dysfunction and reduction of the expected results. Apart from that, a good organizational structure that measures responsibility for results and the system of motivation and rewarding can encourage entrepreneurship and generation of new strategic ideas. The transition from traditional to sustainable management of energy is a big strategic change and inevitably leads to the need to review the existing organizational structure and redefine its parts and hierarchical levels (Golušin, Munitlak-Ivanović, & Redžepagić, 2013).

Development of organizational structure of the company

Growth and development of a company do not follow a straight rising line, but undergo stages of evolution and revolution, characterized by various problems and managerial efforts. All companies spend more or less energy during a certain period of time, and in many companies the cost of energy is quite high. Besides, globally accepted legislation imposes the need to implement the sustainable management of energy as soon as possible and in a more efficient way. However, the company has to justify its business to fulfill its mission and vision and survive in the market. Quantitative growth in the size of the company causes qualitative changes so as to make the effects of size possible (economies of scale) and diversification of activities. In the periods of evolution and revolution (Fig. 5.3) that characterize the interventions in the field of organizational structure, crises occur, or stages come in succession, which require changes in organizational structure and management style. Ichak Adizes described the process of growth and aging of organization through different stages of the life cycle, as follows: courtship, baby, go-go, adolescence, top-form, stability of the aristocracy, early bureaucracy, bureaucracy, and death (Adizes, 2016).

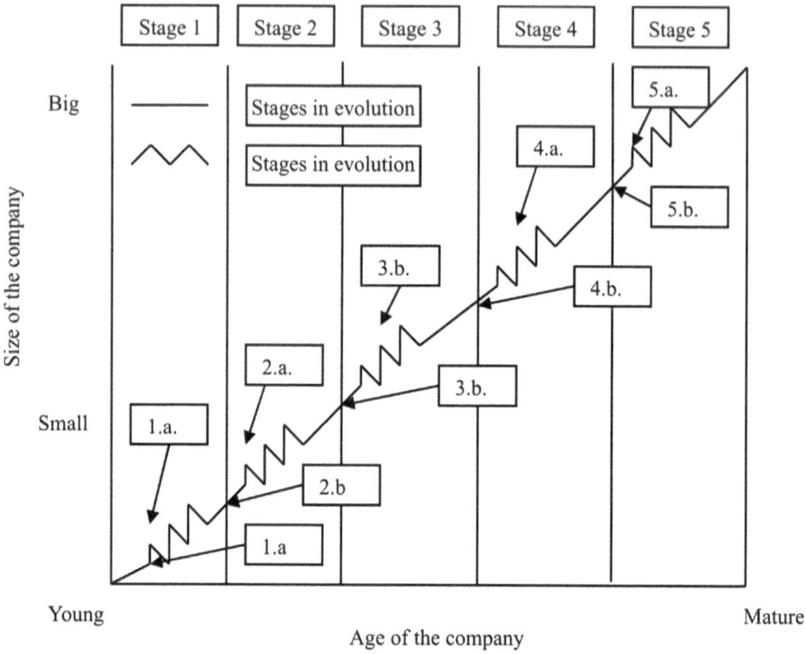

Fig. 5.3 Stages in the development of an organization.

This interpretation can serve as a starting point for companies that plan to switch to sustainable energy management and to keep such a way of doing business, where potential changes are anticipated in terms of the size of the company over time. During this time, each company works within a smaller or larger capacity and therefore consumes lower or greater amounts of energy. It is therefore necessary that management of the company be aware of the changes that can inevitably be expected.

Many factors in the external and internal environment affect the form of organizational structure that will be accepted by the company as a basis for implementing the strategy. As mass production causes excessive greenhouse gas emissions, the gases that cause the "greenhouse" effect, it is of paramount importance to take into consideration economic and ecological implications of the strategy implementation when choosing the appropriate organizational structure (Nie, Li, & Huang, 2016).

In this regard, it is important to determine the range of management and control. There are two possibilities: a high (deep) organizational structure with many hierarchical levels between top management and direct supervisors, or a flat (shallow) organizational structure with few hierarchical levels between the highest and lowest levels of management. Since a good organizational structure should provide effective management and control, the choice between shallow and deep organizational structure must be geared toward gaining competitive advantage, organizational units, and business functions. The selection of the depth of the organizational structure is mostly determined by the energy intensity of the activity that the company deals with. For companies that spend large amounts of energy, a deeper organizational structure is preferred, although one has to be careful here, because too many hierarchical levels often impose burdens and slow down the system of decision-making.

Stages in creating organizational structure

Because of the relationship between strategy and organizational culture, the following stages can be defined in the process of creating an organizational structure that will effectively support the implementation of the sustainable management of energy:

- Identification of the key functions and activities for implementation of the strategy of sustainable management of energy, by which it is necessary first of all to define the activities and responsibilities, rights and responsibilities, and the chain of responsibilities;
- Understanding the relationship between activities in the company's value chain, by which all activities in the process of sustainable energy management are integrated into the existing system of values of the company;
- Grouping activities in organizational units, by which special attention should be paid to defining every organizational unit as an entity with its own peculiarities of production and/or usage of energy, which can be described and set independently from other organizational units;
- Determining the degree of autonomy and independence of each organizational unit, which further enhances the possibility to determine energy characteristics for each organizational unit; and
- Ensuring a system of coordination of activities of organizational units as a natural consequence of interconnectivity of all organizational units, i.e., natural and technologically conditioned cycle of changing the matter and energy.

A particular issue relates to the organizational structure in cases of internationalization of business, because in that way a company significantly expands the scope of its activities and its influence, thus accepting greater social responsibility in the field of sustainable energy management. In these circumstances, it is important to determine the relationships between the peak (the center) and organizational parts that are located in different countries (Martin, Muûls, de Preux, & Wagner, 2012). Apart from the production and geographical criteria, in order to identify the organizational structure, it is necessary to determine the areas of communication between the center and the affiliations in countries and the degree of their independence. In this sense, the company chooses one of the following forms:

- Division structure, which is based on creating divisions as units for coordinating and managing international operations in a given country;
- Global production structure, which is based on the division of the company's organizational units according to groups of products that operate as international units. This form is used when a company is diversified and has significant market share; and
- Matrix structure, which implies that the central unit of the company is responsible for production planning, while geographical divisions are responsible for local sales, distribution, and administration.

Defining corporate culture

When it comes to corporate culture, organizational structure is viewed as a framework and corporate culture is an invisible hand, i.e., a catalyst that provides harmonious functioning of the company. It is usually expressed through a system of values, decision-making, and the dominant model of organizational behavior (Weerts, Vermeulen, & Withes, 2018). This concept is based on the mutual interaction of seven key categories, known as the concept of 7-S. The 7-S framework includes seven categories: the strategy, structure, systems, human resources (staff), shared values, style, and skills. This concept emphasizes the necessity of putting corporate culture in the function of variables such as strategy, system of business flows and information, organizational structure, management style, personnel structure, and the adopted system of values and skills. As shown in Fig. 5.4, an accepted system of values puts corporate culture in the middle, where it permeates all other components and drives the behavior of the company.

Environmentally friendly and socially responsible business is an imperative of contemporary business. A responsible attitude toward energy, energy resources, and energy consumption must be integrated into a business.

Some authors describe corporate culture as the style of doing business, organizational climate, the ego of the company, or beliefs and ideas shared

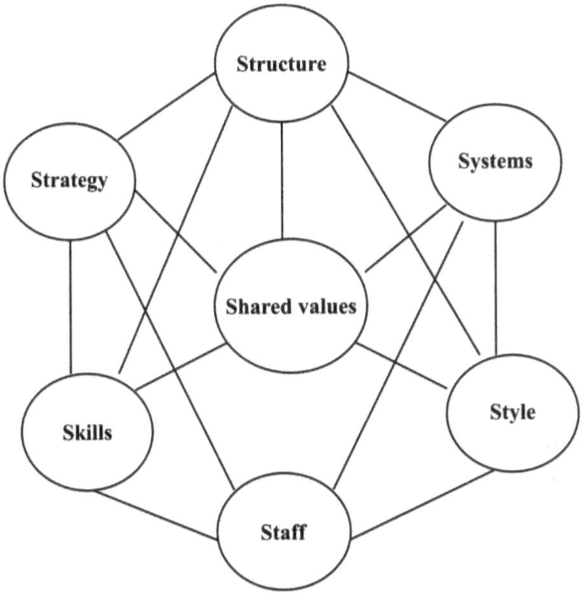

Fig. 5.4 7-S framework.

by the employees. It is compared with the personal identity of individuals, as an accepted system of values, norms, and expectations that determine the way in which companies resolve concrete business situations. Corporate culture gives direction and sense of action; it helps employees identify priorities and properly implement formal activities. In practice, it determines the way in which employees and managers of companies treat customers and suppliers, how they respond to competition, and perform all other activities. Corporate culture is vital if it is characterized by the following: unity, innovation, a clear understanding of company goals, a high level of loyalty, and responsibility to the company or customer orientation (Flamholtz & Randle, 2011).

Since corporate culture consists of value attitudes and norms of behavior according to which employees behave, it is hardly visible, unlike other components on which the implementation of the strategy is grounded. Thus, implementation of the strategy involves harmonization of the structure, competence, allocation of resources, methods of selecting and rewarding human resources, activities of business functions, policies, procedures, and so on.

In the implementation of sustainable energy management, building appropriate business culture is a matter of great importance. In this way, the opinions and attitudes of both management and employees are changed, and the economic performance becomes the only measure of the value of the company. Accepting the need for a responsible attitude toward energy is a slow process that requires an exceptional commitment from management, as well as education and motivation of all employees, along with the participation of external stakeholders (Kegan, Laskow Lahey, Miller, Fleming, & Helsing, 2016).

From this point of view, the concept developed by McKinsey Consultants Comp (Jharotia, 2019) is useful. This concept is based on the mutual interaction of seven key categories, known as the concept of 7-S. The 7-S framework includes seven categories: the strategy, structure, systems, human resources (staff), shared values, style, and skills. This concept emphasizes the necessity of putting corporate culture in the function of variables such as strategy, system of business flows and information, organizational structure, management style, personnel structure, and the adopted system of values and skills. As shown in Fig. 5.4, an accepted system of values puts corporate culture in the middle, where it permeates all other components and drives the behavior of the company.

The strategy of sustainable energy management is directly or indirectly connected to all other categories. The strongest interaction is with the category of staff and with the category of shared values, which is the result of the need to implement the strategy by people who share the same values. Sustainable management of energy requires huge changes in understanding and accepting new values of doing business by employees. Doing business just to gain economic benefits is no longer an imperative in business.

In addition, there is a strong interaction with the categories of skills and structure. Energy management requires certain skills and engagement of people at different structural levels. Companies that are major producers or consumers of energy are also the companies with complex organizational structures. Integration of sustainable energy management into all levels of an organizational structure is often difficult but essential.

Transformation of goals and allocation of resources

As the strategy defines the ways of realization of the mission and goals of the company, an important issue for the implementation of the strategy is the transformation of the new goals into current tasks performed by certain business functions and organizational units of the company. In order to implement the strategy of sustainable energy management in practice, resource solutions related to space, time, and carriers are required. In fact, this part of implementing the strategy is about transforming strategic plans into tactical operations, which refer to shorter time intervals, with fewer alternatives and concrete structured problems. For this reason, the implementation of the strategy requires spatial, temporal, and subject-compliant action plans for one accounting period.

Energy management is inherently a long process and is connected with numerous changes, which significantly extends the time needed for implementation and assessment of positive or negative effects.

A spatially harmonized plan of activities implies that all activities related to sustainable energy management must be precisely determined in terms of space. First of all, it is necessary to determine the parameters of the physical environment in which strategic activities will be conducted. Depending on the activities conducted by the company, spatial determination involves defining activities performed on particular equipment, tools, parts of equipment, connections, important devices, assembly lines, plants, etc. Spatial compliance has to be accompanied by a natural or technological energy flow in a company. For each defined area, it is necessary to determine the activities related to the energy in that segment, which actually ensures measuring of energy consumption, implementation of the measures to save energy, and monitoring the amounts of the waste energy.

Harmonization of time related to the strategy of sustainable energy management includes its integration into the company's overall strategic plans. The implementation of the strategy for sustainable energy management often requires changing the existing goals. One of the goals of every company is certainly to achieve the biggest possible volume of production with the lowest possible costs. In the case of implementing a strategy of sustainable energy management, this is often impossible in the beginning, because the transition to a new method of management is associated with increased initial costs. Therefore, the time for resource allocation is extremely important. The capital has to move from the place where it was previously assigned to the place where it will be used for implementation of the strategy of sustainable energy management. Often this means that the company has to change existing goals and give up some specific goals. Most often companies give up on goals that require the use of large amounts of energy and/or production of energy-inefficient products (Wong, Wong, & Boon-itt, 2020).

Operationalization of the strategy may relate to the operationalization of new business operations or improving or maintaining the current level of operations. Thus, project plans related to the implementation of complex and unique activities may be needed, and the standard plans are used for regular and recurring business situations. The introduction of the strategy of sustainable energy management is usually a major strategic change and requires significant changes. A company will usually create a new product and business portfolio and keep the products and operations that are acceptable in terms of the new strategy but will gradually reduce and eventually abolish energy inappropriate aspects of business. All this requires good planning and significant investments and, certainly, enough time. Changes should be planned and not jeopardize the business ability of the company (Yatsenko & Hritonenko, 2020). The transition to a fully sustainable business in terms of energy can be a significant change and as such it should be carefully approached.

Implementation of the strategy refers not only to the identification and implementation of projects but also to the planning of their performance through the network action plan with clear allocation of resources. Investment programs are developed and selected, as well as the selection of those who would implement the program and models for carrying out programs and projects. In practice, it is not just about standards and familiar and repetitive projects, but it is also about the implementation of previously started but radically new business activities. For these reasons, apart from the *know-how* in the field of project management, it is necessary to build in the *project culture*. The project culture requires teamwork, precise responsibility, and parameters for the evaluation of workers involved, their charisma, and other skills.

As implementation of the strategy requires dimensioning the role and involvement of all business functions, it is important to establish priorities in selecting individual activities and creating resource allocation. Although the company sets the issue of energy management as a priority, it is necessary to implement it in line with other business activities. These issues can refer to the company as a whole or be treated as issues relating to a certain level of organization. The implementation of the chosen strategy can assume the following (Practical guide for implementing an energy management system, 2015):

- The need to increase total resources, where the company can realize the investment in purchasing new energy efficient equipment or its parts or to provide energy from alternative sources;
- The change in the existing allocation of resources, which happens if a company does not have enough money but still wants to do certain adjustments. In this case, the company transfers the existing equity from

one place to another. It ignores the previously defined objectives, and priority is given to solving the problem of unsustainable waste of energy. In this case, the parts of the company are being threatened or even deliberately destroyed in order to create new, energy-friendly parts; and

- The reduction of the required resources and thereby their redirection to an area where resources are needed, as the case that is theoretically possible but not recorded in strategies of energy management because they essentially require increased investments.

It is evident that a certain allocation of resources is usually applied in the company for the needs of sustainable energy management. Allocation of resources can be implemented in various degrees, depending on the degree of centralization and intensity of change, as shown in Fig. 5.5.

When the implementation of the strategy does not require a significant change in resources, the allocation of resources will be done as it was done before. This is often the case in companies that need to implement low-level changes, and most often these are companies that are engaged in activities with low energy consumption. In this case, there is no need for the allocation of resources. It is usually enough to monitor them precisely and control them.

If the management style is based on the partial involvement of individual units in the allocation of resources, it is possible to change the allocation of resources instead of the current practice ("formula"), and the decision will be based on cost analysis and alternative use of resources. This situation is called *conditional bargaining*, and it is specific to companies in which no significant correction of the existing energy flows is necessary, but there are some difficulties in implementation because the system of decision-making is centralized (Schopf & Voss, 2019).

When the decision-making is decentralized, changes in the volume and allocation of resources can occur through *free negotiation*, especially if

Degree of centralization of the direction	Low	High
High	"Formula"	Imposed priority
Medium	Conditioned negotiation	Conditioned bidding
Low	Free negotiation	Open competition

Fig. 5.5 Allocation of resources at the company level.

implementation of the strategy does not require the excessive use of resources. In this case, changes in the field of energy are small, but employees are free to implement them in their own organizational units on their own. This model has its advantages, because employees in certain business units often best know the specific characteristics of that unit, are informed about energy consumption and the reasons for inefficient spending, and can see opportunities for saving energy (Mertzains, 2020).

In the case when the implementation of the strategy involves changing the size and priorities of the engagement of resources (in the case of the development of intensive strategy), and centralization is strong, top management *imposes priorities,* and based on this, the redistribution of resources is done. This is very often the case with companies that transfer to sustainable management of energy. In this case, energy management becomes a priority selected by the management of the company. This imperative becomes a priority of business and is transferred to employees and the requirements they have to meet. Such an autocratic approach can show good results in a short time, but should later be abandoned (Tagliapietra et al., 2019).

In order to stimulate organizational units and consider the environment realistically, these values should be a part of the budgeting process, which is often called *conditioned bidding* (and in practice is very similar to conditioned negotiation), where organizational units of the company apply for funds by bidding. In practice, this means that the transition to sustainable energy management is not an absolute priority, and its implementation is conditioned by the ability of the organizational unit to fight for the necessary funds. Concretely the organizational part of the company that plans to introduce changes in energy management is seeking to mobilize funds from the existing budget by providing the proof that the change is needed and sustainable in the long term. This model is especially acceptable for big companies that operate in many locations.

If an *open competition* between organizational units is chosen, organizational units are encouraged to do better. Such an approach respects the autonomy of organizational units and the need for self-responsibility and achievement of specific organizational units. It is therefore important to estimate the importance of the activities in the company's value chain, respecting the diversity of the focus of some business functions. In this way, conflicts or bias in the distribution (allocation) of resources can be avoided. The above model is acceptable for companies with a high level of corporate culture and is not recommended in the first stages of transition to sustainable energy management (Balmus, Huber, & Ploner, 2020).

Managing the process of implementing sustainable energy management

The essence of the implementation of the strategy of sustainable energy management is the realization of all predicted activities. In order to make the implementation of the strategy of sustainable management of energy efficient, it is necessary to design the organizational structure, allocate resources and related procedures, create an effective system of coordination and information system, make the selection of human resources for key positions, and create a motivating system of rewards.

The implementation of the strategy of sustainable energy management mainly depends on people, and for their motivation and encouragement for action, it is necessary that there is *a leader* who has charisma, knowledge, information, and the authority to reward and punish. However, at the same time, there should be *a manager* who has conceptual and technical knowledge and who through the processes of communication and coordination with average people achieves above average results. The management process generally requires the coexistence of the leader as the creator of the business idea; the business portfolio of the company; and the manager who through planning, organizing, managing, and controlling manages the work processes. The leader deals with challenges and encourages employees to work efficiently or find the right ideas. He is long-term-oriented and motivated to innovate and ask questions (Kaplan, Cortina, Ruark, LaPort, & Nicolaides, 2014). On the other hand, the manager is focused on knowledge that refers to effective management of work processes in order to realize the vision of the company, i.e., he is oriented toward the realization of ideas in the right way.

By vertical differentiation, a division is made in the number of hierarchical levels and by horizontal differentiation, related activities are grouped into productive units and business functions to achieve efficiency. Therefore, coordination is necessary (vertical and horizontal) for different levels of activity as one of the central issues on which the success in implementing the strategy depends.

Coordination of the strategy

Coordination can be defined as the process of integrating the goals and activities of differentiated business units to effectively implement the strategy. If differentiation is more prominent, coordination is more complicated, because it is difficult to find a link and the ability to achieve synergy among the parts of the company. This situation is most common in companies that do business with several major dislocated organizational units, and the need to transfer to sustainable management of energy is not considered equally important in all parts of the company.

At the same time, it is difficult to preserve the coherence of activities and flexibility that often do not go together. A significant and relevant source for acquiring the competitive advantage often lies in the possibility of the transfer of knowledge from one unit to another or combining activities that are used in several business units in a diversified company (Hamdoun, Jabbour, & Othman, 2018). It is necessary to transfer expert knowledge in the field of sustainable energy management from one part of the company to another. In practice, it is often justified to consider the engagement of external experts and consultants, who will monitor the implementation of the strategy of sustainable energy management.

The danger of differentiation may occur due to possible conflicts and bias of management in relation to the organizational unit. The task of coordination is to detect potential conflicts and misunderstandings and manage them in a timely manner and, if possible, avoid such situations. Conflicts often arise because different organizational units understand the basic strategic orientation differently, use the same resources, and have different views on their role and importance in the realization of strategic orientation (Hideo, Yoto, & Kazuhiko, 2020). Very often, the company adopts the implementation of sustainable energy management, but it is implemented differently in terms of its effectiveness in different parts. In order to identify the conflict areas and situations in a timely manner, top management of the company can:

- Rely on basic management techniques;
- Increase cooperation potentials; and
- Reduce coordination needs.

Rely on basic management techniques. In this type of management, coordination is realized through a single chain of command with clearly defined hierarchical relationships between the units and the flow of information adjusted accordingly. Procedures and stipulating rules can be defined so they provide independent and rapid implementation of the tasks of organizational units. To a certain level, unity can be provided through insisting on general objectives and corporate culture that will provide links for coordination.

Increase cooperation potentials. This type of management is used when due to the diversification of activities and shallow organizational structure it is not adequate to rely on prescribing rules or a hierarchy of relationships.

In this case, there is a need to increase the potentials of coordination by using different measures and mechanisms by (Dos Santos, Figueiredo, & Pais, 2020):

- Encouraging direct contacts between leaders to address common issues, in order to familiarize all with the particular features of the sustainable energy management;
- Appointing a liaison officer between the various entities so that each unit has a person to coordinate its work with other organizational units;
- Formulating an "ad hoc" group, comprising representatives of concerned units that will coordinate joint activities, which is especially acceptable for introducing the method of sustainable management, as this represents a big change and requires a special attitude of the involved;
- Establishing a permanent committee responsible for the coordination of joint work, resulting in the possibility that the strategy of sustainable energy management will be implemented in the whole company simultaneously;
- Appointing a deputy general manager or a special manager for the integration of cooperation among organizational units as a necessity in large companies that spend significant amounts of energy;
- Shaping matrix organization in which activities of business functions and organizational units are cross-integrated; and
- Establishing a complete division at the top of hierarchy of the company for coordination of the activities of organizational units.

Reduce coordination needs. This approach insists on the reduction of the need for coordination, by allowing some organizational units greater flexibility in creating their programs and transforming them into independent entities. This approach is acceptable in companies with a high degree of organizational culture, because it means freedom for each organizational unit to select and implement changes in the field of energy independently.

Resistance to change

The introduction of sustainable energy management can often be a big strategic change that requires intensive changes in the company's processes, but also in the attitudes of all employees. This major strategic change always encounters smaller or greater resistance that may last for either a short or long period of time, but it can also completely prevent the required changes. Therefore, for effective coordination in implementing the strategy, it is very important to anticipate the active and passive resistance to change (Lines, Sullivan, Smithwick, & Mischung, 2015). The reasons for resistance can be numerous, including jeopardizing individual interests, misunderstandings, differences in understanding the directions of change, inertia, time period when the change is made, etc. Hence, in order to achieve efficient coordination and effectively manage the process of implementation, it is necessary to make an extra effort in the field of communication, clarification, education, involvement in proposing changes, setting up systems of rewards, etc.

As a method of effective implementation of sustainable strategy of energy management that reduces the potential resistance, the concept of management by objectives can be used. It is a concept in which the involvement of executives in the process of defining goals provides greater motivation and responsibility, defines strategies and objectives in a more realistic way, provides a stronger cooperation in implementing the strategy, etc. The goals and tasks are set in a separate agreement that deals with relations between junior and senior executives and includes (Groves, 2019):

- Sphere of activity for each supervisor;
- Desired goals of activities (short-term and long-term);
- Ways of measuring implementation of goals;
- Limitations and freedoms that each executive has in achieving stipulated goals; and
- Assistance executives can expect from authorities and other organizational units.

Top management of the company has the dominant role in managing the process of implementation of the strategy of sustainable energy management. These bodies interpret and symbolize the strategy, initiate change, and are responsible for effectiveness and efficiency in the execution of the defined strategy. Centralized management is the management in which the top hierarchy defines the strategy for each organizational unit. Participatory management is applicable to cases where the process includes the heads of organizational units. In practice, there are combinations of methods and management styles that depend on effort in terms of process planning and implementation of the strategy.

Selection of managers

The strategy of sustainable management of energy requires engagement of appropriate managers who will deal with all phases of management of the process. The first question concerns the selection of managers. The selection of managers depends on the abilities and needs of the company. Also, the choice of the staff is important, i.e., whether to choose from the staff already employed with the company or to recruit new employees. The advantages of selecting the existing workforce is that they are already familiar with other employees, business practices, work environment, their characteristics are well known to other staff members, they have established contacts with customers, suppliers, and colleagues and thus show their loyalty to the company. Disadvantages may arise due to the inability to accept change, workload, personal attitude, and the like. Using people who have gained experience in implementing similar changes is an advantage, since they desire to prove themselves in a new environment and are not overburdened by inherited relations toward associates. The disadvantages of this solution include the high cost of attracting such experts, the difficulty in finding the right person, the negative impact on loyalty of other employees, etc. The implementation of sustainable energy management is a major strategic change and involvement of experts outside the company is almost inevitable, but it is certainly desirable to create mixed teams of external experts, full-time employees, and experienced managers and engineers.

When selecting key managers, two important aspects should be taken into account: (1) changes necessary for implementation of the new strategy and (2) past performance of companies (Hadad, Keren, & Laslo, 2013). Taking into account these two coordinates, the following matrix shown in Fig. 5.6 can be obtained.

		Past performance	
		Effective	Ineffective
Neccesary changes for the implementation of new strategy	Many	*"Selective mixture"* In-company managers are combined with the new managers (outsiders) through promotion and transfer and provision of knowledge and experience	*"The turn"* Hiring new managers (outsiders) so as to provide new ideas, skills, enthusiasm, and motives for success
	Few	*"Stability"* Orientation towards internal promotion of in-company executives so as to develop, keep, and reward their managerial flair	*"Reorientation"* Hiring managers (outsiders) so as to surpass weaknesses along with promotion and reallocation of the existing executives

Fig. 5.6 Approaches to selection of managers.

Selective mixture describes the situation when the implementation of strategy requires a lot of changes, but those changes are not initiated by the previous performances. Since the previous performance was good, it is a good choice to rely on the strength of existing management, and selective outsourcing of external managers can ensure changes implied by the new strategy. This situation arises when the company has effectively run the process of energy management and gained valuable experience.

Stability is a situation that arises when the implementation of the strategy does not require a radical change, i.e., it is intended to preserve current performance and relies on the promotion of existing managers in order to build customer loyalty and further develop management talent. This situation is typical for companies that are usually small consumers of energy, and sustainable energy management is not a big change.

The turn is characterized by a situation where it is necessary to intensify efforts to implement defined strategies due to ineffective operations. In this case, bringing in new managers becomes more important in order to provide fresh ideas, new skills, motives, and enthusiasm. In most companies, sustainable management of energy is well planned, but it is not implemented, mainly because of the absence of commitment at top management level. In such a situation, bringing in new managers is quite justified, because the old ones usually are big potential source of resistance to change.

Reorientation characterizes the situation where the implementation of the strategy does not require a lot of changes and previous performance was bad. In this situation, it is best to bring in external managers (outsiders) to help with critical issues in the field of business functions or reallocate existing managers.

Hiring of new managers and their participation in the activities of the company and the implementation of strategies for sustainable energy management largely depend on the company's ability to pay for their engagement and other benefits offered to new managers. Therefore, the structure of the reward system is very important for the process of implementation of the strategy. These are sensitive issues of great importance for the success of the implementation of the strategy. Within these considerations, the form of giving incentives should be defined (profit sharing, stocks, cash, etc.), as well as techniques and the basis for calculation, bonuses, rights of the managers based on hierarchical position, the frequency of payments, consequences of risk, and other related issues.

Control and audit of the strategy

The final stage of the process of strategic management is the control and audit of the implementation of the strategy of sustainable management of energy. This phase should provide a more realistic basis for implementation of the strategy and should enable the timely implementation of strategic options, as well as help to promptly take any corrective action.

Timing of the control

According to Herni Fayol, control of an undertaking consists of seeing that everything is being carried out in accordance with the plan that has been adopted, the orders that have been given, and the principles that have been laid down. Its object is to point out mistakes in order that they may be rectified and prevented from recurring (Fayol, 2013). Control of the implementation of the strategy of sustainable energy management should be conducted according to the plan and in the time determined. In this regard, and from the point of view of time, control may be conducted during the following periods:

- After formulating a global strategy but before its implementation;
- During the implementation of the strategy; and
- After implementation of the strategy.

The first control, which refers to the formulation of the strategy of sustainable management of energy, is aimed at checking the compatibility of the strategy with the mission, resources, weaknesses, and strengths of the company, external conditions, corporate culture, and time interval. Control during the implementation has the task of timely pointing to possible deviations during the execution in relation to the plan. Finally, the control after the implementation of the strategy is the basis for a change and a new plan.

The basic purpose of control and review of the strategy is to timely identify and evaluate possible deviation from the plan and take corrective actions (Diaz, Ocampo-Martinez, & Olaru, 2020). If the process of planning is the first in the chronology of strategic management, when decisions are made about the objectives, policies, strategies, plans, and programs, then the mechanism of control that timely qualifies the flow of these activities is completely logical. In this sense, control should meet two purposes (Tsamenyi, Sahadev, & Quiao, 2011):

- To discipline those who plan and execute planned activities; and
- To contribute to the realization of the process of planning in the next period.

It is easy to see the correlation between planning and control. Planning and control are linked through the implementation of the strategy, and it makes no sense to plan if that plan will not be controlled. On the other hand, it is not possible to control activities that were not planned previously (Adler & Chen, 2011). These two processes are mutually encouraged and derive one from another. It is necessary to determine the basic parameters and the scope of control as follows:

- Subject of control.
- System of control.
- Control and audit of the strategy.
- Control of the strategy in the environment.

Subjects of control

From the point of view of the dimension and the subject of control, as shown in Fig. 5.7, it is possible to differentiate strategic, managerial, and operational control. The implementation of the strategy involves the activities of all hierarchical levels in the company: from the top of the company through strategic business units and business functions. This significantly determines the system of control.

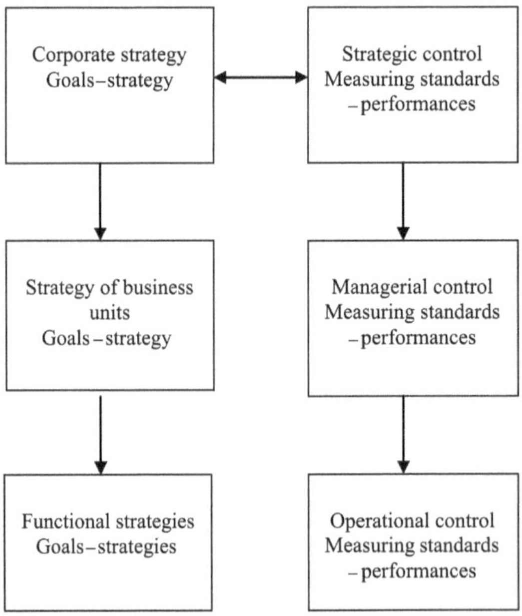

Fig. 5.7 The relationship between strategy and control.

On the other hand, the types and standards of control depend on the structure of activity, degree of decentralization, management style, and reward. Typically, strategic control refers to control of the implementation of basic strategic orientation within the competence of top management, which is of great importance to the company as a whole. This control is based on long-term, broad, and quantitative standards. The shareholders' assembly, supervisory board, and the external audit (auditing performed by certified organizations) assess the skills of top management. The rating of management refers to their ability to adequately evaluate the general and branch environment, identify critical factors of business success (hot spots), and choose the appropriate overall strategy. Fig. 5.7 shows that management control is related to the control of the implementation of strategies at the level of individual business or programs, or whether it is in the domain of top management and heads of business units. In general, the control of the operation involves the control of the implementation of strategy of business functions and control of individuals and groups in accordance with the prescribed standards and procedures.

The type of system of control and efficiency of implementation depend on management style and the system of reward, or set of procedures and standards for monitoring. This is necessary because it is hard to distinguish when strategic, managerial, and operational controls start and when they finish, due to the interconnection of hierarchical levels in the implementation of the strategy. To understand this sensitive area in a realistic way, it is important to determine the standards and procedures more accurately. As synthetic indicators related to strategic control, it is possible to measure the level of productivity, product quality, work morale of employees, earnings per share, the dynamics of introducing new products, market share, the costs of obtaining resources, etc. (Sljivic, Skorup, & Vukadinovic, 2015).

The degree of success at the strategic level depends on the success of organizational units and business functions that are integrated and differentiated at the same time in the overall process of implementation of the strategy. For this reason, in addition to strategic control, operational control is elaborated, which directly identifies performances toward carriers and the places of strategy implementation. Operational control is performed more frequently than strategic control and performance indicators are more specific. The very process of defining standards, measuring achievement, comparison with standards, qualifying, and if necessary, taking corrective actions, is adapted to the planned implementation of the strategy. Fig. 5.8 shows the importance of the system of budgeting, which is considered the forerunner of strategic planning.

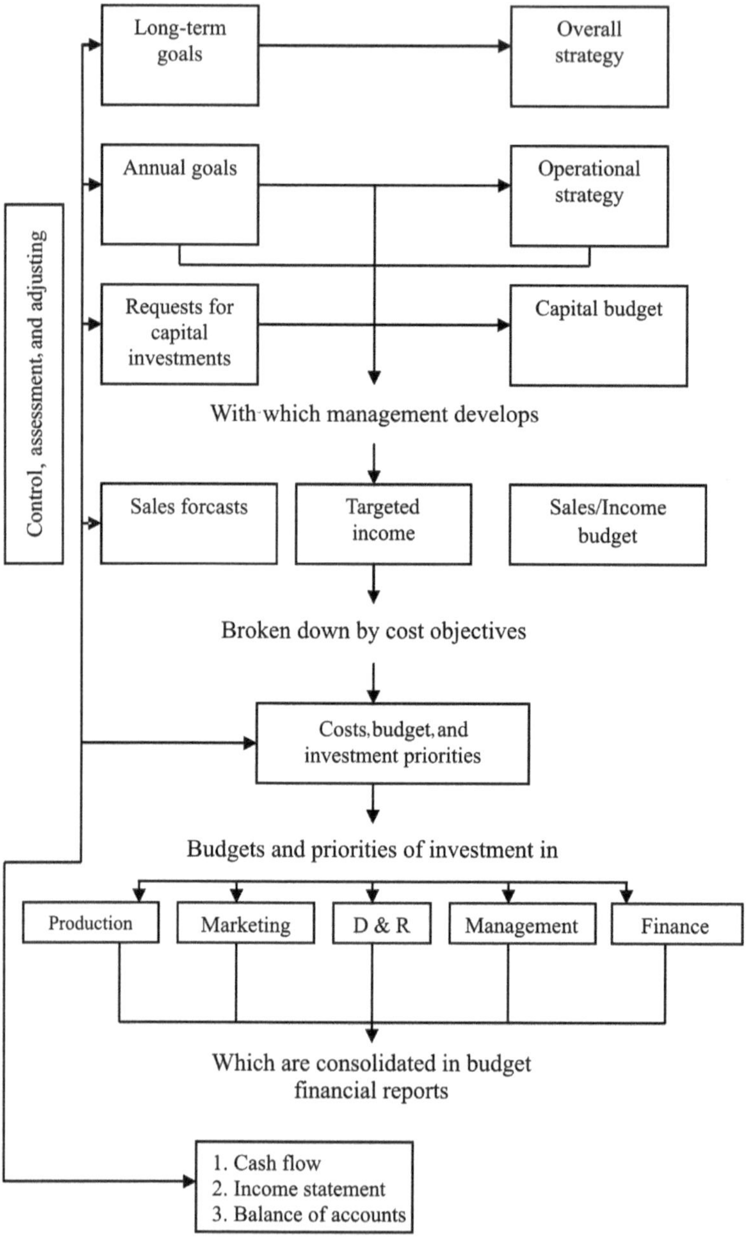

Fig. 5.8 Budget system for control of implementation of the strategy of sustainable management of energy.

The system of budgeting sets the network for planned allocation of resources and coordination and control of their use for their intended purposes by the carriers of certain activities. The budgeting system establishes standards that bind together human, physical, and financial resources for the accounting (planned) period. This system is of great importance because it determines the investment dynamics and priorities. By monitoring the indicators of the fulfillment of the plan (budget) of sales, it is possible to estimate the effectiveness in implementing the strategy and to timely take corrective action if the practice requires.

The plan of expenditures, if structured according to functional areas, makes it possible to monitor the implementation of each activity according to the prescribed standards, for example, sales per employee, ratio of sales and investments in marketing activities, increase in productivity, etc. A capital budget has similar characteristics and provides the ability to control types, levels, and areas of investing the capital.

System of control

The necessity of reliable and comprehensive control, along with the existence of different standards, has influenced the development of a number of different control systems.

Objective criteria, such as stock prices, return on invested capital, and the level of transfer prices, are the characteristics of *market control*. *Output control* is a control that takes into account the fulfillment of the division goals, business functions, and an individual's goals and requires a careful approach because the compliance of the quality and quantity of goals must also be taken into account. *Bureaucratic control* is more difficult to implement in practice, because it requires the definition of standards, procedures, and budgets for carrying out certain activities. *Group control* should identify the system of values, i.e., corporate culture, which directly depends on toleration and interpretation of certain activities as well as the ability of the employees to develop and maintain good communication (Hui & Fatt, 2007).

Control and audit of the strategy

There is a great similarity between control and audit, and these two procedures should not be separated if we speak about the strategic component of control. While control determines the relationship between the achievement and the plan, the audit finds out whether the plan has been achieved and fulfilled in accordance with the possibilities. The control has a framework that entails the reality of the plan and the intention to keep business within the domain of the plan. On the other hand, apart from the plan, the audit takes into account the new challenges and opportunities that result from changes in the external and internal environment and suggests a possible change of plan. The audit, along with the control, is the element of planning, but it is based on understanding the trends in the environment and new management ideas.

The audit of the strategy of sustainable management of energy is based on two important assumptions: (1) predicting as a basis of input of planning is limited and (2) business planning depends on the style of management (Aghdam, Salehi, & Ghaemi, 2018). In the first case, the audit will be based on multivariate (contingent) planning, i.e., the planning will include multiple scenarios and plans. In the second case, the emphasis is on the change of management style as a direct consequence of new leaders and managers, leading to the need to review business philosophy, content, and the planning process itself. The market environment is very unpredictable: the longer the time horizon of the future, the less reliable the forecasting. This indicates that the audit becomes increasingly important when it comes to strategic control.

An audit can be defined as a comprehensive, independent, systematic, and periodic review of environment, strategy, objectives, and activities of companies or business units with the aim of identifying problems and opportunities, in order to propose a plan of action that will improve the business performance of companies. The relationship between audit and control is tight, because an adequate audit requires, apart from making predictions, taking into consideration the results given by the system of control of the previous activities.

The process of strategic management involves iteratively linked steps that are aimed at ensuring efficient and effective business. On the other hand, the process of making strategic decisions usually requires long-term engagement of the company and during that time circumstances in which the company makes changes (Chofreh et al., 2020). Accordingly, assessment, control, and audit are needed in order to determine whether the strategy as a result of strategic management is in accordance, above or below expectations or planning (Schulze & Heidenreich, 2015). For these reasons, the audit itself involves several areas, as follows:

- Evaluation of internal harmonization of strategic goals with other goals of the company, because the implementation of the strategy of sustainable management of energy often has direct or indirect effects on achieving other goals;

- Evaluation of the quality of the process of analysis, which was the basis for the development of the strategy, as a way to determine if the previous assessment of the condition in terms of energy in the company was correct;
- Evaluation of the very contents of the strategy (direction, pace, and method of growth), because the strategy of sustainable management of energy usually takes longer and usually goes at an uneven pace;
- Evaluation of the opportunity of the company to achieve a formulated strategy, since regardless of orientation, it can happen that a company is unable to move to sustainable energy management in a way and to the extent envisioned by that company; and
- Evaluation of results brought about by the implementation of the strategy, which is achieved by measuring concrete outcome results—the most important are savings in energy and increased energy efficiency.

Control of the strategy in the environment

It is necessary to distinguish between an unstable and stable environment and adjust techniques on which the control of strategy is based. If the environment is stable, it is relatively easy to predict, and the role of the control is to observe the progress of the company. In an unstable environment, when predictions are hard to make, the task of the control is to assess the assumptions, goals, and components of the strategy. Two concepts of control derive from this attitude: (1) control of the current trend in the company and (2) control of changes undertaken by the company (Otley & Soin, 2014).

In case of the control of the current trend, which is applied in a relatively stable environment, the aim is to analyze where the company is in relation to the trend. Strategic control is intended to determine whether, despite the changing environment, assumptions on which the strategy is formulated are still valid. The strategy of sustainable energy management is carried out primarily on the basis of the strong influence of the environment, and in that sense, it can be expected that the requirements would be the same and even emphasized. This largely guarantees justification of the change brought about by the new strategy.

Corrective actions are taken based on the control if the implementation deviates from the plan. In this case, previously defined general strategies expand. This form of control is based on creating special centers of responsibility, identifying success factors, and identifying performance standards for general strategies (Strauß & Zecher, 2015).

These are units for which the domain of responsibility and authority is determined, as well as standards against which achievement of the formulated strategy of the company will be monitored. In this sense, it is usually the creation of *revenue centers, cost centers, profit centers,* and *investment centers*. In the case of *cost centers*, the costs of production and supply are controlled. *Revenue centers* control realization of the planned performance in terms of marketing and sales. *Profit centers* control management of revenues, costs, and profit, thus they are considered a fairly comprehensive form of control. The most comprehensive form of control is the control by *investment centers*, which involves, apart from revenues, costs, and profit, investment management and assets management.

Control by the creation of centers of responsibility is a subsequent control of costs, revenues, profits, and investments in the field of energy management. In contrast to this concept, *the control based on factors of business success* is proactively oriented, because it examines the factors that contributed to (non-)success of the strategy and assesses whether the formulated strategy is still valid (Berens, 2014).

The domain of the previous forms of control that starts from the premise that there are no radical changes in the environment is not enough to propose the strategy through assessment and audit, if radical changes occur. In this case, it is necessary to apply *control based on the changes undertaken by the company*. In such cases it is necessary to rely on control of innovative undertakings that aim to audit the entire process of strategic management. This concept of control should help in volatile business conditions in the environment, which are characterized by irreversible changes. This form of control is predominantly proactively oriented and involves a much wider scope of testing and is also based on more sophisticated techniques: the subtle techniques of evaluating the components of the company, the possibility of combining for the needs of synergy, the concept of simulation, the concept of managing important strategic issues, methods, and scenarios, i.e., contingency planning, etc.

Such an audit has the characteristics of business restructuring, as it aims at removing the gap between the current and potential state and requires a diagnosis, a prognosis, and proposed changes (Pusnik, Al-Mansour, Sucic, & Gubina, 2016). However, the audit differs from restructuring because the restructuring involves intervention to remove the already evident crisis, and revision is the result of the proactive attitude that it is possible to achieve more and better business operations. Restructuring is a subsequent, often belated response to changes in the environment, and the audit is the expression of proactive action, thanks to periodic and timely review. If the audit is well implemented, restructuring will not be necessary.

As the basic characteristic of a strategy is to adequately regulate relations with the environment, evaluations and audits are identified with the strategic control. Therefore, this stage of the process of strategic management is the responsibility of top management of the company.

Case study

Prioritizing critical success factors for sustainable energy sector in China: A DEMATEL approach, Energy Strategy Reviews, Volume 35, 2021, https://www.sciencedirect.com/science/article/pii/S2211467X21000213.

References

Addis, B., Capone, A., Carello, G., Gianoli, L. G., & Sansò, B. (2016). Energy management in communication networks: A journey through modeling and optimization glasses. *Computer Communications*, *91–92*, 76–94.

Adizes, I. (2016). *Mastering change—Introduction to organizational therapy (revised & updated edition)*. Adizes Institute Publications.

Adler, P. S., & Chen, C. X. (2011). Combining creativity and control: Understanding individual motivation in large-scale collaborative creativity. *Accounting, Organizations and Society*, *36*, 63–85.

Aghdam, P. H., Salehi, J., & Ghaemi, S. (2018). Contingency based energy management of multi-microgrid based distribution network. *Sustainable Cities and Society*, *41*, 265–274.

Ahmad, M., Jan, I., Jabeen, G., & Alvarado, R. (2021). Does energy-industry investment drive economic performance in regional China: Implications for sustainable development. *Sustainable Production and Consumption*, *27*, 176–192.

Balmus, T., Huber, J., & Ploner, M. (2020). More competition in delegated portfolio management: A win-win situation? An experimental analysis. *Journal of Economics Behavior & Organization*, *178*, 777–800.

Berens, W. (2014). A broad field of management control. *Journal of Management Control*, *24*, 299–300.

Brown, P., Pham, H., & Sivabalan, P. (2020). Automation and management control in dynamic environments: Managing organisational flexibility and energy efficiency in service sectors. *The British Accounting Review*, *52*, 100840.

Casco-Vidrean, L., Niemann, J., Baidoc, R., & Pisla, A. (2018). Change management aspects in solar energy implementation. *Procedia—Social and Behavioral Science*, *238*, 432–441.

Chen, B., Xiong, R., Li, H., Sun, Q., & Yang, L. (2019). Pathways for sustainable energy transition. *Journal of Cleaner Production*, *228*, 1564–1571.

Chofreh, A. G., Goni, F. A., Klemeš, J. J., Moosavi, S., Davoudi, M., & Zeinalnezhad, M. (2020). Covid-19 shock: Development of strategic management framework for global energy. *Renewable and Sustainable Energy Reviews*, *110643*.

Diaz, J. L., Ocampo-Martinez, C., & Olaru, S. (2020). Dual mode control strategy for the energy efficiency of complex and flexible manufacturing systems. *Journal of Manufacturing Systems*, *56*, 104–116.

Dos Santos, N. R., Figueiredo, C., & Pais, L. (2020). Development and validation of the organizational cooperation questionnaire. *European Review of Applied Psychology, 70*, 100555.

Fadnavis, S., Najarzadeh, A., & Badurdeen, F. (2020). An assessment of organizational culture traits impacting problem solving for lean transformation. *Procedia Manufacturing, 48*, 31–42.

Farham, H., Mohammadian, L., Alipour, H., & Pouladi, J. (2019). Energy procurement of large industrial consumer via interval optimization approach considering peak demand management. *Sustainable Cities and Society, 46*, 101421.

Fayol, H. (2013). *General and industrial management*. Martino Publishing.

Flamholtz, E. G., & Randle, Y. (2011). *Corporate culture: The ultimate strategic asset*. Stanford Business Books.

Frate, G. D., Ferrari, L., Desideri, U., Sorbi, F., Bosi, N., & Lazzari, M. (2018). Energy and economic savings through a plant supervised management in large-scale commercial activities. *Applied Thermal Engineering, 141*, 269–279.

Golušin, M., Munitlak-Ivanović, O., & Redžepagić, S. (2013). Transition from traditional to sustainable energy development in the region of Western Balkans—Current level and requirements. *Applied Energy, 101*, 182–191.

Graves, M. L., Sarkis, J., & Gold, N. (2019). Employee pro-environmental behavior in Russia: The roles of top management commitment, managerial leadership, and employee motives. *Resources, Conservation and Recycling*, 54–64.

Groves, K. S. (2019). Confronting an inconvenient truth: Developing succession management capabilities for the inevitable loss of executive talent. *Organizational Dynamics, 48*, 100668.

Hadad, J., Keren, B., & Laslo, Z. (2013). A decision-making support system module for project manager selection according to past performance. *International Journal of Project Management, 31*, 532–541.

Hamdoun, M., Jabbour, C. J. C., & Othman, H. B. (2018). Knowledge transfer and organizational innovation: Impacts of quality and environmental management. *Journal of Cleaner Production, 193*, 759–770.

Hideo, H., Yoto, K., & Kazuhiko, T. (2020). A study of conflict solving tactics and culture in integrated organizations through interviews and text mining. *Procedia Computer Science, 176*, 1960–1970.

Higuchi, Y., Mhede, E. P., & Sonobe, T. (2019). Short- and medium-run impacts of management training: An experiment in Tanzania. *World Development, 114*, 220–236.

Hui, L. T., & Fatt, Q. K. (2007). Strategic organizational conditions for risks reduction and earnings management: A combined strategy and auditing paradigm. *Accounting Forum, 31*, 179–201.

Isensee, C., Teuteberg, F., Griese, K. M., & Topi, C. (2020). The relationship between organizational culture, sustainability, and digitalization in SMEs: A systematic review. *Journal of Cleaner Production, 275*, 122944.

Javied, T., & Franke, R. J. (2015). Implementing energy management system to increase energy efficiency in manufacturing companies. *Procedia CIRP, 26*, 156–161.

Jharotia, A. K. (2019). McKinsey's 7S model for academic libraries. In *Conference: McKinsey 7S Model: Innovation growth, sustainability and scalability of business, New Delhi, India.*

Kaplan, S., Cortina, J., Ruark, G., LaPort, K., & Nicolaides, V. (2014). The role of organizational leaders in employee emotion management: A theoretical model. *The Leadership Quarterly, 25*, 563–580.

Kegan, R., Laskow Lahey, L., Miller, M. L., Fleming, A., & Helsing, D. (2016). *An everyone culture: Becoming a deliberately developmental organization*. Harvard Business Review Press.

König, W. (2020). Energy efficiency in industrial organizations—A cultural-institutional framework of decision making. *Energy Research & Social Science, 60*, 101314.

Kuo, T. C., & Smith, S. (2018). A systematic review of technologies involving eco-innovation for enterprises moving towards sustainability. *Journal of Cleaner Production, 192*, 207–220.

Lima, E. S., McMahon, P., & Cabral Seixas Costa, A. P. (2020). Establishing the relationship between asset management and business performance. *International Journal of Production Economics*, 107937.

Lines, B. C., Sullivan, K. T., Smithwick, J. B., & Mischung, J. (2015). Overcoming resistance to change in engineering and construction: Change management factors for owner organizations. *International Journal of Project Management, 33*, 1170–1179.

Liu, C., Fang, J., & Xie, R. (2021). Energy policy and corporate financial performance: Evidence from China's 11th five-year plan. *Energy Economics, 93*, 105030.

Martin, R., Muûls, M., de Preux, L. B., & Wagner, U. J. (2012). Anatomy of a paradox: Management practices, organizational structure and energy efficiency. *Journal of Environmental Economics and Management, 63*, 208–223.

Mertzains, C. (2020). Financial supervision structure, decentralized decision-making and financing constraints. *Journal of Economic Behavior and Organization, 174*, 13–37.

Nie, S., Li, Y. P., & Huang, C. L. (2016). Risk management of energy system for identifying optimal power mix with financial-cost minimization and environmental-impact mitigation under uncertainty. *Energy Economics, 61*, 313–329.

Noboa, E., & Upham, P. (2018). Energy policy and transdisciplinary transition management arenas in illiberal democracies: A conceptual framework. *Energy Research & Social Science, 46*, 114–124.

Otley, D., & Soin, K. (2014). *Management control and uncertainty*. London: Palgrave Macmillan.

Practical guide for implementing an energy management system. (2015). Vienna: United Nation Industrial Development Organization.

Pusnik, M., Al-Mansour, F., Sucic, B., & Gubina, A. F. (2016). Gap analysis of industrial energy management systems in Slovenia. *Energy, 108*, 41–49.

Schneider, B. (2020). People management in work organizations: Fifty years of learnings. *Organizational Dynamics*, 100789.

Schopf, N., & Voss, A. (2019). Bargaining over natural resources: Governments between environmental organizations and extraction firms. *Journal of Environmental Economics and Management, 97*, 208–240.

Schulze, M., & Heidenreich, S. (2015). Linking energy-related strategic flexibility and energy efficiency—The mediating role of management control systems choice. *Journal of Cleaner Production, 140*, 1504–1513.

Si, D. K., Wang, Y., & Kong, D. (2020). Employee incentives and energy firms' innovation: Evidence from China. *Energy, 212*, 118673.

Sljivic, S., Skorup, S., & Vukadinovic, P. (2015). Management control in modern organizations. *International Review, 3-4*, 39–49.

Strauß, E., & Zecher, C. (2015). Management control systems: A review. *Journal of Management Control, 23*, 233–268.

Tagliapietra, S., Zachmann, G., Edenhofer, O., Glachant, J. M., Linares, P., & Loeschel, A. (2019). The European union energy transition: Key priorities for the next five years. *Energy Policy, 132*, 950–954.

Tsamenyi, M., Sahadev, S., & Quiao, Z. S. (2011). The relationship between business strategy, management control systems and performance: Evidence from China. *Advances in Accounting, 27*, 193–203.

Weerts, K., Vermeulen, W., & Withes, S. (2018). On corporate sustainability integration research: Analyzing corporate leaders' experiences and academic learnings from an organizational culture perspective. *Journal of Cleaner Production, 203*, 1201–1215.

Wong, W. C., Wong, C. Y., & Boon-itt, S. (2020). Environmental management systems, practices and outcomes: Differences in resource allocation between small and large firms. *International Journal of Production Economics*, *228*, 107743.
Xia, L., Wei, J., Gao, S., & Ma, B. (2020). Promoting corporate sustainability through sustainable resource management: A hybrid decision-making approach incorporating social media data. *Environmental Impact Assessment Review*, *85*, 106459.
Yatsenko, Y., & Hritonenko, N. (2020). Optimal asset replacement: Profit maximization under varying technology. *International Journal of Production Economics*, *228*, 107670.

CHAPTER 6

Control of sustainable energy management implementation

Contents

Elements of strategic control

Control may be conditionally considered as the last step in the process of sustainable energy management. Its essence is to compare the achieved with the planned, identify possible deviations, and undertake corrective activities. Control is hence only in theory perceived as the last strategic step, while, essentially speaking, it represents a starting point for making necessary corrections to the energy management process, planning it again, and performing it with no interruptions and with constant improvements. Control therefore focuses on goals stemming from an enterprise's mission. Definition of the basis that can be used for contrasting is one of the main issues (Doerr, 2018). A well-conceived control offers information on negative deviations and their removal at early stages, prior to the conclusion of envisioned activities. Control is a system that has to be explicitly and clearly set, based on its contrasting processes and unambiguous principles (Niven & Lamorte, 2016).

https://doi.org/10.1016/B978-0-12-821086-4.00009-X

Control system

Control may be defined as "a process ensuring that implemented activities match the envisioned ones" (Qemaji, 2016). Objectives and their implementation are defined in the planning stage. Control should measure fulfillment of a plan, i.e., the level of achieving set goals, establishing deviations, and proposing corrective activities.

Control is a systematic effort, and envisioned goals become controlling standards: defining the feedback and comparing real results with previously defined standards, identifying size, and importance of deviations (Wodtke, 2021). Control should also deal with taking appropriate actions that use all available resources to reach objectives in an effective and efficient manner. Such a definition implies that the control process consists of four stages, as shown in Fig. 6.1.

The four control process stages include defining standards and result-measuring methods, results measuring, contrasting achieved results with standards, and undertaking corrective measures (Falke, 2021). A strategic energy management control system has a number of specificities.

Defining standards and result-measuring methods

As has already been underlined, planned objectives are most often taken for standards and basic elements of control. Objectives are defined and at the same time result-measuring methods and responsibility of entities (managers most often) are established (O'Farrell, 2020). If standards are not set, managers may respond to small deviations or, more seriously, fail to react to significant and alarming deviations. The motivation function and comparison "threshold" function are two important functions of standards in the control

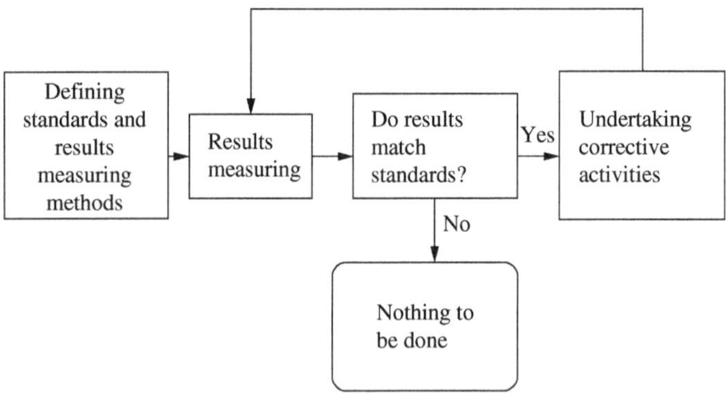

Fig. 6.1 Control process.

process. Objectives will give better results and control will be efficient, provided objectives are motivational and not too low (Grove, 2015). Most employees feel better and are more satisfied if they manage to respond to challenges and be successful in execution of high standards. However, if they are too high and cannot be objectively carried out, it would be counterproductive and bring about motivation decline. In this case, it is better to spend more time and set up new standards that objectively fit business terms and conditions (The Arbinger Institute, 2018).

Standards should be comprehensible for those that execute them, which is one of the main prerequisites. It is necessary to articulate adequate result measurements so that an objective can become a control standard and result-measuring methods reliable. For example, when discussing services, the waiting time may be set up as a control standard. In manufacturing, measures should synthesize quality control indicators and production and financial indicators as well (Jacobs, Berry, Whybark, & Vollmann, 2018).

Results measurement

Unlike the previous stage of control, results measurement is a repetitive activity, the frequency of which depends on the activity being measured. For example, the level of greenhouse gases a chemical or metallurgy plant emits in the environment should be measured on a daily basis, while the growth rate of an enterprise may be measured once a year. Selection of activities to be measured should be established in line with the management principle of *exception based*. A manager who is responsible for a certain process should only be informed about the activities if significant deviations from standards are recorded. In this case, the manager should react so that activities are adjusted and an adequate solution that would result in compliance with a standard for a certain activity is found (Willink, 2020).

Most systems function according to the feedback principle. This is particularly relevant to result-measurement systems, since by measuring results, obtained information is communicated to decision-makers.

Comparison of achieved results with standards

This stage of control is, in practice, the easiest step of the controlling process, since if results are in line and at the level of set standards, no intervention or corrective activity is needed. Setting standards in an adequate way and measuring results is much more complex than determining whether there are any deviations from such standards (Nise, 2019).

Undertaking corrective activities

If the previous stage proved there are deviations from standards, the control process stage is needed. The corrective activity itself might be simple and involve changing only one activity, or it may include a number of activities or alternations of previously defined standards. The control process is a dynamic process, and if managers apply it afterward, i.e., *ex ante*, and follow it only through its results, it can be said that they only follow results but fail to "exercise control." Control should be *exercised,* so that results get closer to standards in a constructive manner before a concrete activity is undertaken (Badiry & Agustiady, 2021).

Control parameters

Frequency, control subject and object, and quantity and quality of information provided through implementation of the control are the main control parameters in the process of sustainable energy management.

The *frequency of control* depends on a number of factors and may be defined at the very beginning, at the planning stage of energy management process. However, changes in frequency of control quite often take place during the process itself. In most cases, frequency of control is under the influence of the following factors (Chaisson, 2022):

- The activity of an enterprise. Controls will be much more frequent in companies that are large energy users. In such cases, both theory and practice urge for everyday controls of implementation of the energy management process, which primarily refers to following energy consumption levels and recording relevant factors. In the course of time, having gained insight into certain regularities in energy consumption, these controls might be reduced; however, they should still be performed at least once a month;
- The range of sustainable energy management the organization wants to carry out has an impact on frequency of control. Sustainable energy management may be conducted in a different range, from partial and symbolic up to comprehensive and full. Naturally, implementation of partial enhancements does not require frequent controls since these measures are often determined only once in the observed period (for example, once a year) with effects visible following the introduction. Implementation of comprehensive sustainable energy management is an extremely complex and long-lasting process requiring the

introduction of frequent, precise, and, in a special way, planned controls, and parameters under control may in this case be numerous and varied;

- Experience in implementation of the management process as a factor that largely and quite often defines frequency of control, despite being based on immeasurable influence. Namely, companies with certain experience in the implementation of a management system in different areas (not only energy management) may believe their experience allows for control to be performed less often since they may be reassured that experience overcomes mistakes resulting from insufficient and imprecise controls. On the other hand, companies with less experience in implementation of the management process very often fail to attach any importance to the control itself, not even in the planning phase; rather they have mainly been devoted to the implementation. In other cases, insufficiently experienced companies may plan for extremely rigorous and frequent control that may burden the system with a great quantity of information that is often useless, and important information may go unnoticed or be insufficiently reconsidered. Introduction of a sustainable energy management process is in most cases a huge change, and companies quite often fail to understand that the sustainable management approach differs greatly from traditional approaches. It is thus recommended due to the complexity of change and insufficient knowledge of sustainability mechanisms that companies have objective frequency of controls regardless of experience. Application of other companies' experiences (benchmarking) could give good guidelines in this area;
- The degree of changes coming from the outside, as a result of the fact that the sustainable energy management process, has been carried out for years. It very likely that during this period, certain changes will take place in internal and external environments that may influence the implementation process and control results as well. The most common changes impacting control standards are those pertaining to adoption of certain legal and international obligations and agreements, significant changes in prices of energy substances, and internal changes in the structure and liquidity of the company itself; and
- Perception of the listed changes by control stakeholders (subjects). Most frequent controls of the implementation of sustainable energy management process have been carried out in companies feeling those changes might have a huge impact on both company and the environment. Furthermore, some companies often fail to conduct any review

or plan any changes in the control system, which leads to irregular and inaccurate and hence useless results. Less-experienced companies often demonstrate insufficient sensitivity toward changes taking place in the environment that may affect an energy management system. It is therefore important for a company to have an elaborated system of providing responses to changes and the capacity to assess the ongoing potential level of influence of changes on expected results and control of implementation at the same time. Any changes in price of energy substances, for companies that are large energy consumers, have an immediate effect on output parameters and the results of the application of sustainable energy management, so an adequate system for financial aspects assessments has to be provided in this sphere in particular.

Apart from the above, the frequency of control in the sustainable energy management process does not depend on every factor acting individually, but most often on the effect created by their mutual interaction. It is therefore very important to be familiar with the rules pertaining to the factors' mutual relations. Significant changes in the energy market may cause significant consequences if a company is run by insecure and inexperienced management (Filipović, Radovanović, & Lior, 2019). On the other hand, there are companies that spend significant quantities of energy substances but believe the costs may be refunded through product prices, in which case the sustainable energy management process becomes irrelevant and marginalized by both management and employees (Islami, Mullali, & Mustafa, 2018).

The *object of control* is of special interest in the control process of implementation of the sustainable energy management process. This control is based on measurement of certain parameters, which are then contrasted to the values stipulated in the goals planning phase and give a success measure of the implemented strategy.

Identification of an object of control, i.e., control items, first has to be harmonized with previously set objectives. Each objective set in the planning phase has to be measurable, so that it can be contrasted in the control phase. Setting objectives in immeasurable units in sustainable energy management is useless.

Apart from harmonizing control items with previously set objectives, control items have to be selected so they provide data relevant for further measurement in the sustainable energy management process. A control item may be any place at which one may obtain certain data within the company.

However, the following controls are considered necessary for sustainable energy management (Barney, William, & Wayne, 2020):

- Measuring the total energy consumption following a certain period of time defined by the control plan (day, week, months), which provides data on consumption trends (increase, decrease, stagnation) with definition of clear reasons for the spotted trend, aiming at arriving at some regularities;
- Specifying the total energy consumption by sectors in a company, with the comparison of the measured consumption with the planned one, registering all parameters that may have influence on the measured level of consumption (capacity, type of products, technology, season, obstacles);
- Establishing the biggest energy consumption area in a certain time frame, aiming at defining areas suitable for the most significant savings;
- Measuring consumed energy by a product unit following a certain control period and comparing that quantity with the quantity defined as a goal, with the purpose of establishing whether conducted measures lead toward the goal at the desired pace;
- Specifying the quantity of energy obtained from the company's own or other energy renewable resources. Following a certain control period, it is possible to establish the quantity of energy replaced by the energy from renewable resources and compare it with the defined goals. It is possible in this way to establish in a timely manner whether measures have been enough to meet this goal (new technological solutions, use of new energy resources, creating alternative energy resources, etc.). Often, companies set unrealistic goals and envision quick and abundant replacement of traditionally obtained energy with energy from renewable resources; and
- Defining financial effects of sustainable energy management measures implementation is of great interest in the overall process since it could largely define survival of implementation of the process pertaining to transfer to sustainable business. The comprehensive transfer to sustainable energy management principally implies increased costs in the very beginning, so that companies using a lot of energy have to be ready to accept initial negative financial effects. Investing into energy-wise efficient technologies and building one's own capacities for production of energy from alternative sources are financially demanding investments, and the return period is often longer than 5 years. Company should expect the positive effects only after a certain period of time.

Defined by application of adequate economic measures for specifying justifiability of investments. Therefore financial reports are one of the control items, monitoring general and incomes and expenses situation in the company, since investments for energy sustainable business are inseparable from other financial business indicators. In some cases, current business results are sufficient to compensate for initial negative financial effects resulting from larger investments in energy solutions. However, quite often positive financial effects cannot be resolved without adequate help from additional investments by foreign investors, special funds, or banks.

The above list proves that energy quantity, efficiency, and financial indicators are the most common control subjects. A combination of these indicators may provide insight into the given circumstances and conduct of the overall system, on the basis of which one may conclude whether the company is moving in the sustainable energy management direction or not. Timely control of the given indicators leaves enough time for review of goals and redefinition of the strategy, provided measured activities prove to be unsatisfactory. Fig. 6.2

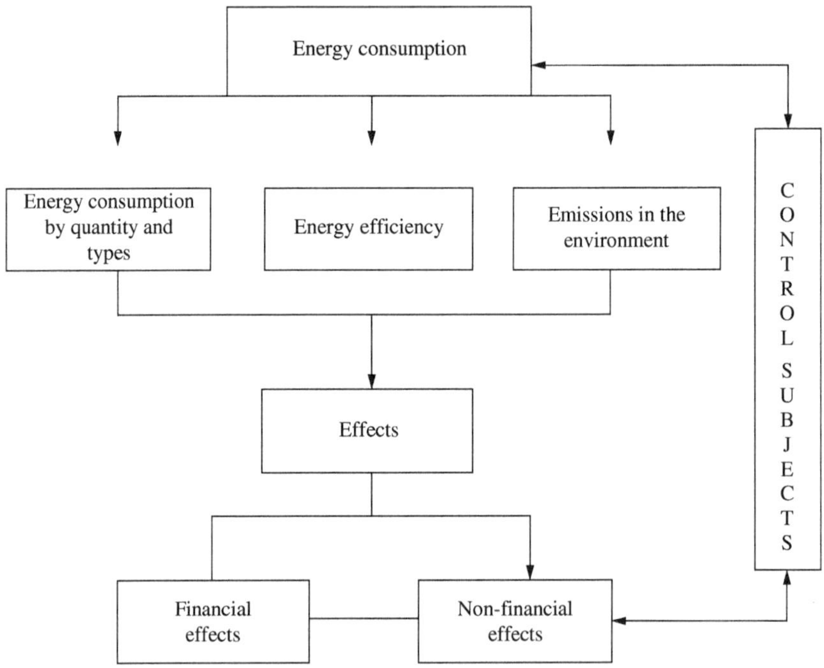

Fig. 6.2 Interrelation among control items in the process of sustainable energy management.

shows the relationships among consumption, energy efficiency, emissions in the environment, and financial effects as main control subjects.

Control of sustainable energy management is a particularly complex process that does not actually have an end result. Control is the process based on measuring the realization of objectives, as well as their effects, but control is performed continuously and does not end after the measurement of certain effects. The essence of control lies in the fact that all measured values are reduced to a single measure, which is suitable for conducting the required analysis, mathematical operations, and comparison with targets. The most appropriate unit for expressing all subjects of control is money (Lee & Cheng, 2016).

Energy consumption is monitored and measured on several grounds, depending on whether the consumption is measured in household consumption, the company, or any particular country, but is most often expressed through consumption by quantity and types of energy sources. All values can be expressed in a certain monetary value. In addition, power consumption can be expressed by certain parameters of energy efficiency that can be expressed in different units, but the most appropriate is the unit of income (Wang, Lou, Fang, & Tan, 2021). Control of emissions into the environment can be expressed in specific measurement units depending on the type of contaminants, but these values can be converted into a certain value in money that takes into account the amount of penalty to be paid if there is a certain overpollution, or as saving (income) that occurs if the pollution is within the allowed limits. Also, at this point, the sum attained on behalf of the premium for the reduction of pollution can be expressed in money (Rosado & Levinson, 2019).

The controller is an issue of special interest that is often not given enough attention. More specifically, companies often approach the process of energy management sporadically, and if there is a total commitment to solving this problem, the issue of entities that initiate and control the whole process and are therefore responsible for conducting of the control cannot be solved in a way that could be considered satisfactory. At this point, it should be noted that the main aim of control is the provision of sufficient information to indicate the correctness of or discrepancies in the whole process. Information obtained by control is used in making important decisions in the process of further planning. Therefore, the issue of control is of special interest (Gopinath, Kumar, Joshua, & Srinivas, 2020).

In all processes of management, the issue of responsibility and control is particularly sensitive because it is usually coupled with the already established system of governance that may or may not be adequate when it comes to the process of sustainable energy management. Therefore, control over this process shares the fate of all other processes of control that are going on in the company, and the subjects who carry it out are most often persons or groups that are in charge of all other areas of control.

The presented process of sustainable management, with its numerous peculiarities, long process, and especially the initial negative financial effects, requires a special approach in the control phase, which therefore requires specially oriented subjects of control. In the optimal case, to implement adequate control, a person or persons should be engaged, having the following characteristics (Michelucci, De Marco, & Tanda, 2016):

- Professional competence, which includes education, skills, and abilities necessary to monitor and control processes such as sustainable energy management, is the first prerequisite in this process. People must have the necessary theoretical and practical knowledge in the field of energy and sustainable development, as well as basic knowledge in management science. It is necessary that these persons are able to identify problems, to distinguish between important, less relevant and irrelevant events in the system, and to react accordingly. It is expected that the person in charge of control is capable of independent and team work.
- Legal competence related to the position that the person or persons responsible for the control of the process of sustainable energy management already take in a particular company. Usually, these are people who are in high management positions, which normally have insight into control of all the processes of management. On the one hand, it is desirable that the sustainable energy management is integrated in other management processes in the company, but it is not good if, because of that, energy management becomes marginalized in some way. The process of sustainable energy management is complex and requires full engagement of people who cannot have other responsibilities.

From the above, it can be concluded that in order to implement the stage of control, it is necessary to engage people who were engaged in the process of planning the strategy of sustainable energy management (Bull & Janda, 2018). Moreover, these people should have been engaged in creating the basic objectives. Apart from that, in companies that consume more energy, the process of transition to the new energy management should be managed by a person who has only that task to do.

Preferably the process should be controlled by a person working in the company, because such a person is aware of the technological processes and the ways of doing business in general, although there are no restrictions regarding the hiring of expert consultants (Da Silva Gonçalves & Dos Santos, 2019). Although outsourcing can bring many advantages, because the people who are engaged as consultants are very experienced professionals familiar with good or poor business practices in this sphere, and are especially objective because they have no stake in the internal interests of the company, which each management process requires.

The amount of information that can and must be gathered by controlling the implementation of the process of sustainable energy management can greatly vary. Too much information in most cases cannot provide a clear picture of the implementation of the strategy of sustainable energy management, because in most cases it overloads the control system and increases the possibility of error. An insufficient amount of information, on the other hand, does not provide enough information on implementing the strategy, and thus it is not possible to make the right decisions on further actions. The optimal amount of information is a parameter that is passed in accordance with experience and in accordance with the specific features of the company in which control is performed and the ability of the system of control that is in use.

Quality of information is a more important parameter of control and indicates the reliability of information or its ability to present the real situation. Measurable information is usually the information of the greatest degree of quality, and it is most frequently obtained in the energy sector, so the quality of information in this case does not present a particular problem provided that the measuring instruments are adequate and precise and readings are regularly made.

Frequency of control depends on the activities of the company and control models. The choice of subjects and frequency of control along with the definition of the basis for comparison are the main problems of control implementation. As the modern environment is characterized by rapid and intense change, it becomes more difficult to set a basis for comparison. Response of the company to changes often increases the degree of diversification and decentralization. On the other hand, decentralization brings new demands. These requirements refer to the coordination and determination of responsibility for decisions with a greater number of participants. In order to more easily respond to the consequences of this change, it is necessary to continually test and timely modify the previously issued decisions.

Controlling financial effects of implementation

The process of sustainable energy management is complex, time consuming, and demanding on several grounds. One of the key problems that slows down the process of transitioning from traditional to sustainable energy management is exactly the problem of planning, monitoring, and evaluation of financial effects of the application of the stipulated strategy (Podbregar et al., 2020). For companies involved in energy-intensive and demanding activities, these problems are severe and therefore require special attention and the development of acceptable methods for assessment. Great difficulties in practice often cause uncertainties in the field of cost-effectiveness and problems in monitoring financial effects during the implementation of sustainable energy management.

In most cases, the transition to sustainable energy management involves significant financial investment in equipment, resources, training, implementation of a series of standards, and engagement of experts and consultants (Kamyab, Klemeš, Fan, & Lee, 2020). All this represents a significant cost to the company. In addition, in the transition to sustainable energy management in the initial phase delays occur most commonly, as well as incompatibility of equipment and people, the problems of energy supply, and the like, which inevitably lead to loss of value of production, thus decreasing revenue. Therefore, the investment in sustainable ways of using energy requires careful monitoring of financial parameters (Elie, Granier, & Rigot, 2021).

To assess feasibility of investment in sustainable energy management, certain methodology should be used that provides consistency in assessment of various investments, thus allowing testing and comparison of indicators over several years (Jin & Hu, 2022). The basic algorithm of estimation of profitability of investment in sustainable energy management involves the following activities:

- Assessment of the developing capacities of the company—which requires real estimation of the current business performance and analysis of the development strategy, new markets, changes in the process of work, estimation of the risk in the environment, and all other economic, financial, social, technical, and technological and other environmental conditions;
- Estimation of current status and the necessary changes—necessary to analyze conditions in the field of energy management in the company, and above all assess in detail the parameters of energy supply (quantity, type, price, conditions given by the supplier) to determine the way in

which energy is consumed (type, quantity, energy efficiency, energy waste) and based on that set a list of priorities that require investment in order to transfer to a sustainable energy management;

- Analysis of procurement options—if the company finds out it is necessary to implement certain changes or invest in new sources of energy, it should estimate opportunities for procurement of necessary equipment for carrying out such an enterprise, because the markets in which the equipment can be found are sometimes too distant and prices and quality are different;
- Estimation of the value of the necessary investment—having chosen the type and quality of the necessary equipment, the total value of investment should be estimated, which provides insight into the total amount of money necessary for carrying out certain investment activity;
- Determining sources of funding—investments in sustainable forms of energy management often require engagement of bank capital, so it is necessary to analyze bank offers and special funds that provide finance for this purpose;
- Estimation of future income—transition to sustainable energy management is often connected with certain problems in the initial stage that inevitably lead to reducing the value of production and thus reducing income that should help companies survive, pay its liabilities, and settle loan installments;
- Estimation of future expenditures—transition to sustainable energy management often requires investments during several years that increase usual expenditures of the company and, apart from that, the expenditures entail liabilities toward sources of funding;
- Static estimation of the cost-effectiveness of investment—comparison of revenue and expenditures in the representation year of investment (the year in which the full capacity of production has been achieved and liabilities toward sources of funding are still active) can provide insight in cost-effectiveness of the stipulated investment;
- Dynamic estimation of the cost-effectiveness of investment—the first impression of cost-effectiveness of investment is created by financial indicators during the whole period of exploitation of the stipulated investment (which estimates in which year the investment would pay off), method of net current value (by applying the method of discounting), and method of internal rate of return (which estimates if the investment would pay off under certain bank conditions);

- Sensitivity analysis—respects the fact that during the whole period of the project certain changes may occur in the environment that would directly influence financial indicators of return. Therefore, it is necessary to estimate to which extent the investment is able to stand them without jeopardizing liquidity. The following has been estimated: changes in the price of energy raw materials, changes in the price of other input, reduction of revenues, increase of expenditures, and other parameters that are considered to be of special importance; and
- Final financial estimation of investment feasibility—carried out based on previously determined parameters, by which it is necessary to ensure that the investment fulfills all stipulated methods of estimation in order to be considered acceptable.

All investments in the systems and facilities for sustainable energy management are to a certain extent subsidized and supported by the state authorities but are still a huge cost for the company and therefore special attention should be paid to estimating financial aspects.

Controlling nonfinancial effects of implementation

Improvements in the field of energy are the consequence of a long-term unplanned exploitation of natural resources and pollution of the environment caused by their exploitation and consumption. The energy problem has been widely recognized as one of the priorities of sustainable development, and it can be solved by implementation of different activities at all levels. All projects in the field of quality management of sustainable energy development attract the interest of huge groups of stakeholders, such as (Guðlaugsson, Fazeli, Gunnarsdóttir, Davidsdottir, & Stefansson, 2020):

- Companies, basic units of business, and consumers of energy, which get a series of advantages by implementation of adequate methods of management that refer to creating positive image, profit, cost reduction, and provision of certain subsidies;
- Beneficiaries, consumers of goods and services provided by the companies, that want to support energy saving companies, and therefore put pressure on institutions and are willing to ensure a healthy environment for themselves and their descendants;
- Citizens, who as consumers of energy stimulate energy saving and investments in renewable energy sources, thus gathering certain advantages, favorable loans, and tax reductions;

- Legislation, as a necessary framework for functioning of the system, should regulate the sustainable energy management and should be changed depending on the changes in requirements imposed by world integration;
- Investors, who want to direct their capital toward companies that are doing business responsibly toward the environment, and that invest in the projects for saving energy or production of energy from renewable sources, because they know that such companies have a more secure future and more secure capital;
- International communities, who are interested in efficient energy management, since energy stability is one of the key issues in terms of the survival of the world;
- Generations that have shown interest in regulating very complex problems of energy management, regardless of the fact that full effects in this field can be expected in the long term; and
- Future generations, who are the biggest beneficiaries of the changes in the field of production and consumption of energy, and who have the right to have at their disposal the same quantity of resources that the previous generations had at their disposal, which is a key principle of sustainable development, called the principle of intergenerational justice.

Case study

Understanding the effects of energy management practices on renewable energy supply chains: Implications for energy policy in emerging economies. Energy Policy, 118: 418–428. https://doi.org/10.1016/j.enpol.2018.03.043.

References

Badiry, A. B., & Agustiady, T. (2021). *Sustainability: A systems engineering approach to the global grand challenge (analytics and control)* (1st ed.). CRC Press.

Barney, L. C., William, J. K., & Wayne, C. T. (2020). *Guide to energy management*. River Publishers.

Bull, R., & Janda, K. B. (2018). Beyond feedback: Introducing the 'engagement gap' in organizational energy management. *Building Research and Information*, *46*(3), 300–315.

Chaisson, J. (2022). *An introduction to system modeling and control* (1st ed.). Wiley.

Da Silva Gonçalves, V. A., & Dos Santos, F. J. M. H. (2019). Energy management system ISO 50001:2011 and energy management for sustainable development. *Energy Policy*, *133*, 110868.

Doerr, J. (2018). *Measure what matters: OKRs: The simple idea that drives 10x growth* (1st ed.). Penguin.
Elie, L., Granier, C., & Rigot, S. (2021). The different types of renewable energy finance: A bibliometric analysis. *Energy Economics*, *93*, 104997.
Falke, K. (2021). *Lead well: 10 steps to successful and sustainable leadership*. Lioncrest Publishing.
Filipović, S., Radovanović, M., & Lior, N. (2019). What does the sharing economy mean for electric market transitions? A review with sustainability perspectives. *Energy Research & Social Science*, *58*, 101258.
Gopinath, R., Kumar, M., Joshua, P. C., & Srinivas, K. (2020). Energy management using non-intrusive load monitoring techniques—State-of-the-art and future research directions. *Sustainable Cities and Society*, *62*, 102411.
Grove, A. S. (2015). *High output management* (2nd ed.). Vintage.
Guðlaugsson, B., Fazeli, R., Gunnarsdóttir, I., Davidsdottir, B., & Stefansson, G. (2020). Classification of stakeholders of sustainable energy development in Iceland: Utilizing a power-interest matrix and fuzzy logic theory. *Energy for Sustainable Development*, *57*, 168–188.
Islami, X., Mullali, E., & Mustafa, N. (2018). Using management by objectives as a performance appraisal tool for employee satisfaction. *Future Business Journal*, *4*, 94–108.
Jacobs, F. R., Berry, W. E., Whybark, D. C., & Vollmann, T. E. (2018). *Manufacturing planning and control for supply chain management: The CPIM reference* (2nd ed.). McGraw Hill.
Jin, X., & Hu, H. (2022). Research and implementation of smart energy investment and financing system design based on energy mega data mining. *Energy Reports*, *8*, 1226–1235.
Kamyab, H., Klemeš, J. J., Fan, Y. V., & Lee, C. T. (2020). Transition to sustainable energy system for smart cities and industries. *Energy*, *207*, 118104.
Lee, D., & Cheng, C. C. (2016). Energy savings by energy management systems: A review. *Renewable and Sustainable Energy Reviews*, *56*, 760–777.
Michelucci, F. V., De Marco, A., & Tanda, A. (2016). Defining the role of the smart-city manager: An analysis of responsibilities and skills. *Journal of Urban Technology*, *23*(3), 23–42.
Nise, N. S. (2019). *Control systems engineering* (8th ed.). Wiley.
Niven, L. R., & Lamorte, B. (2016). *Objectives and key results: Driving focus, alignment, and engagement with OKRs* (1st ed.). Wiley.
O'Farrell, I. (2020). *SMART objective setting for managers: A roadmap (performance development series)*. Evolution Consulting.
Podbregar, I., Šimić, G., Radovanović, M., Filipović, S., Maletič, D., & Šprajc, P. (2020). The international energy security risk index in sustainable energy and economy transition decision making—A reliability analysis. *Energies*, *13*, 3691.
Qemaji, S. (2016). Notion of strategic control. *European Journal of Interdisciplinary Studies*, *1*(5), 122–130.
Rosado, P. J., & Levinson, R. (2019). Potential benefits of cool walls on residential and commercial buildings across California and the United States: Conserving energy, saving money, and reducing emission of greenhouse gases and air pollutants. *Energy and Buildings*, *199*, 588–607.
The Arbinger Institute. (2018). *Leadership and self-deception: Getting out of the box* (3rd ed.). Berrett-Koehler Publishers.
Wang, Y., Lou, D., Fang, L., & Tan, P. (2021). Energy management and emission control for range extended electric vehicles. *Energy*, *236*, 121370.
Willink, J. (2020). *Leadership strategy and tactics: Field manual*. St. Martin's Press.
Wodtke, C. (2021). *Radical focus: Achieving your most important goals with objectives and key results* (2nd ed.). Cucina Media, LLC.

CHAPTER 7

Strategic priorities of sustainable energy development

Contents

First strategic objective: Exploitation of renewable energy sources

Global development of society in the future will largely depend on the situation in the field of energy. Problems faced by most countries in the world are associated with the provision of energy and environmental protection. Explosion of human population on the earth has caused a constant increase in the demand for energy, particularly electricity. On the other hand, the current structure of primary energy sources cannot meet such an increasing demand for electricity on a global scale. The reason for this also lies in the environmental problems caused by the direct combustion of fossil and

Sustainable Energy Management
https://doi.org/10.1016/B978-0-12-821086-4.00004-0

nuclear fuels the present generation of electricity is based on. In addition, the current dynamics of consumption of fossil fuels will lead to the disappearance of reserves in the near future (Waheed, Sarwar, & Wei, 2019).

The direct consequence of these opposite conditions of production and consumption is a steady increase in electricity prices, which already at the current level create an environmentally and economically justifiable need for the inclusion of alternative energy sources in a global energy development strategy. These energy flows have forced highly developed countries to invest huge capital and employ a large number of experts in the development of alternative renewable energy sources (wind power plants, solar power plants, biomass and biogas power plants, geothermal energy, etc.). As a result of such an investment, the technology industry has adopted and developed technically reliable conversion of some primary renewable sources. In addition, international protocols and commitments to reduce CO_2 emissions and local environmental problems have forced many governments to stimulate, through various subsidies, the construction of environmentally clean power plants using renewable sources of electricity (Jordaan et al., 2019).

These energy flows have forced highly developed countries to invest in the construction of electrical power plants that use wind power and solar power, currently the most promising alternative renewable sources of electricity.

Solar energy

The potential of solar energy

The energy of the sun's rays coming to the earth's surface, i.e., the potentially usable radiation of the sun, is about 1.9×10^8 TWh/year. This energy is approximately 170 times greater than the total energy of coal reserves in the world, and when compared with the energy needs of mankind, it amounts to 1.3×10^8 TWh/year. This means that the solar energy that reaches the earth's surface during only 6 h is sufficient to meet all the needs of the world on an annual basis. In order to get better insight into these data, an average household in some of the most developed countries spends about 10,000 kWh of electricity a year, and it would take about 100,000 years to spend 1 TWh (Kabir, Kumar, Kumar, Adelodun, & Kim, 2019).

About 37% of world energy demand is met by the production of electricity that during 2008 amounted to 17,000 TWh. If this energy is generated by photovoltaic (PV) systems (systems that convert solar energy into electricity) of modest annual power output of 100 kWh/m^2, a surface of

$150 \times 150\,km^2$ would be necessary for the accumulation of solar energy. Much of the absorption surface could be placed on roofs and walls of buildings and would not require additional surface area on the ground (Andrae & Edler, 2018).

The energy of solar radiation is sufficient to produce an average of 1700 kWh of electricity per year per square meter of soil, and the greater the radiation at a site, the higher the generated energy. Tropical regions are more favorable in this respect from other regions with moderate climate. Secondary radiation in Europe is about $1000\,kWh/m^2$, while, by comparison, it is 1800 kWh in the Middle East.

Application of solar energy can be achieved in two ways:

- Converting solar energy into heat; and
- Converting sunlight into electricity.

Solar systems for heat production are used in households, industry, agricultural buildings, and other facilities, that, for example, use large quantities of sanitary water. However, in the last 10 years, PV conversion of solar energy has become the primary industry of solar devices due to the large number of technological advantages in relation to the conversion into heat and due to the rapid development of relevant technologies and their projected abilities.

Production of PV devices has doubled every 2 years, with an average annual increase of 48% since 2002, so this economic sector shows the fastest growth in the world compared with all other branches of energy technology. From an economic point of view, the price of electricity produced from solar energy is continuously decreasing as a result of technological improvements and growth in mass production, while it is expected that fossil fuels will become more expensive in the near future (Crago & Koegler, 2018).

It is now justifiable to encourage the use of solar energy for the production of heat and electricity in the area of households, industry, and some branches of agriculture due to lower investment costs. However, in the long run, the future conversion of solar radiation is in PV technology and its integration with other branches of technology, which is in accordance with the attitudes, plans, and current status in the European Union and other leading economic countries.

The technology of solar PV devices

The direct conversion of solar energy into electricity, the so-called PV effect, was perceived almost two centuries ago, but it was only after the development of quantum theory at the beginning of the 20th century that this phenomenon was explained and understood. The first solar PV cell was developed at Bell Laboratories in 1954. With the development of space exploration, PV cells made of semiconductor silicon quickly became the main source of electricity for satellites, primarily because of its reliability, while the price was of minor importance. The importance of terrestrial use became topical in the world energy crisis of the early 1970s. Today, PV conversion implies a high-technology production of electricity from solar energy.

PV systems consist of modules made of solar cells. PV systems are modular so their strength can be designed for virtually any application. Additional parts to increase the output power are easily adapted to existing PV systems, which is not the case with conventional energy sources such as power plants and nuclear power plants whose economic viability and feasibility require multimegawatt installation. PV modules contain a number of serial- or parallel-connected PV cells used to obtain the desired voltage or current. Solar cells are laminated between two protective layers. On one side, there is a specially tempered glass with low iron content and on the other Tedlar protective plastic material or another layer of glass (Fig. 7.1). In a typical solar module, solar cells are integrated and laminated with the laminating plastic (EVA). Such a laminated PV module is protected from adverse environmental impacts to extend service life. The typical manufacturer's warranty on a PV module is 25 years. PV panels include one or more modules that can be used individually or in groups to form a modular system, together with the

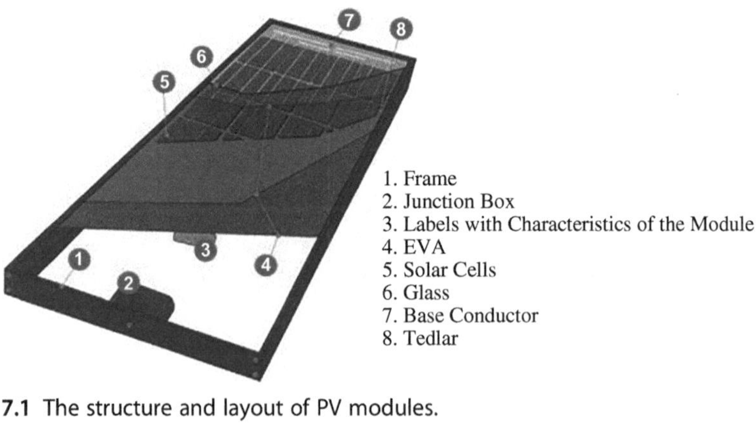

Fig. 7.1 The structure and layout of PV modules.

supporting structures and other necessary components (Fu, Qin, Bai, Yang, & Chen, 2019). Systems can be fixed in a certain position toward the sun or mechanically adjusted continuously toward the sun (adjustable supporting systems).

Latest generation of PV solar modules consist of so-called thin-layer PV cells and panels in which the thickness of the PV material or layer is about 2 μm (micrometers or microns), almost 100 times less than the traditional PV cells made of crystalline silicon. This reduction in the quantity of used semiconductor material reduces the cost per unit area, the price of the generated power (expressed in W), and the cost per installed kWh and increases profitability. However, the total price (balance of costs) of a PV system is still somewhat higher for thin-layer PV modules because of the additional costs of supporting structures required to install these modules and to a lesser extent because of the costs of inverters and connecting to the distribution network. Also, operation and maintenance are more expensive because of the greater surface area they cover in relation to the classical PV modules, so these additional costs must be compensated by the benefits arising from savings of materials. In the past several years, the price, related to larger module surface, has been constantly falling due to substructure optimization, higher module efficiency, and better quality of electrical connections (Park, Lee, & Park, 2018).

The standard components of PV systems are (Mohanty, Muneer, Gago, & Kotak, 2016).

- PV modules;
- Controllers and battery charge regulators;
- Accumulators or batteries;
- Cables and mounting systems; and
- AC/DC power converters—inverters (autonomous and network).

Direct current produced in solar cells or modules leads to the controller through the cable. The primary function of the controller is to prevent overcharging the battery, but there are some other roles depending on the specific application. If the battery is not fully charged, electricity is free to go to the battery, where energy is stored for later use. If the system needs to run devices that work on AC, part of the PV system will also be inverters that convert DC to AC electricity. If the PV system is linked to an electric power distribution network, the so-called special network inverters that allow synchronization of the PV system with the network and restoration of electric power back into the network are used. In these situations, the network is the medium for electric energy storage instead of batteries. It is the most widespread application of PV systems in developed countries today (Kumar & Bhimasingu, 2017).

The excess energy generated in autonomous PV systems during sunny periods is collected in batteries, and some independent operations, for example, for direct pumping of water or running other engines, do not require the use of batteries. The water is pumped when the sun is shining and is directly stored in a reservoir located at a higher level for later pumping through the effects of gravity. Other PV systems convert DC to AC power and inject excess electricity into the electrical distribution network, while taking energy from the grid during the night, when there is no sunlight. This is an example of the operation of PV systems connected with a distribution system network. The three typical configurations of PV systems are the autonomous system, the system connected to the distribution network, and the hybrid system. The autonomous and hybrid systems can be used independently, so they are not connected to the electrical distribution network and are often used in physically remote areas. PV systems connected with electricity distribution networks are one of the ways to create decentralization of the electrical network. Electrical energy is generated by these systems closer to the locations where there is demand, not only by power plants, nuclear power plants, or large hydropower plants. Over time, these systems will reduce the need for increased capacity of transmission and distribution lines (Chaudhary & Rizwan, 2018).

High-temperature solar collectors

In solar thermal power plants, solar radiation through lenses and mirrors is concentrated to achieve higher temperatures. This technique is called a concentrated solar energy. Depending on the temperature of the medium in the solar power plants, appropriate engines are used. Steam turbines are used up to 600°C temperature of the working medium, while gas turbines are used for achieving higher temperatures. Very high temperatures condition the use of different materials and techniques. For temperatures over 1100°C, liquid fluoride salts should be used as the working medium and multistage turbine system, which can be achieved, and the degree of efficiency to 60%. High operating temperatures enable a large savings of water, which is particularly important in deserts where the construction of such plants is expected (Pelay, Luo, Fan, Stitou, & Rood, 2017).

As the storage of thermal energy is cheaper than electricity, solar thermal power plants are usually performed with thermal tanks. This enables the production of electricity at night. Solar power plants can be a reliable source of electricity in locations with good solar radiation. Reliability can provide additional power to fossil fuels (see Fig. 7.2).

Advantages of solar energy exploitation

The advantages of using solar energy are numerous. First, the source of energy is practically inexhaustible. The sun always shines on certain land, and at any time half the planet isn't illuminated by sunlight. Sunlight is free and as such, accessible to all. In addition, regardless of the brightness disparity, there is the possibility of a relatively accurate prediction of the amount of radiation and thus of the amount of energy that can be obtained in this way. The great advantage of using solar energy is that the energy obtained in this manner can be stored in batteries and used when needed (Ahmed, Sreeram, Mishra, & Arif, 2020).

Disadvantages of solar energy exploitation

The main disadvantage of the production and use of solar energy is its price. Namely, the production, installation, and maintenance of the equipment needed to convert solar energy into electricity are relatively expensive. The production of energy from the sun is a few times more expensive than the production from other alternative sources. In addition, the energy

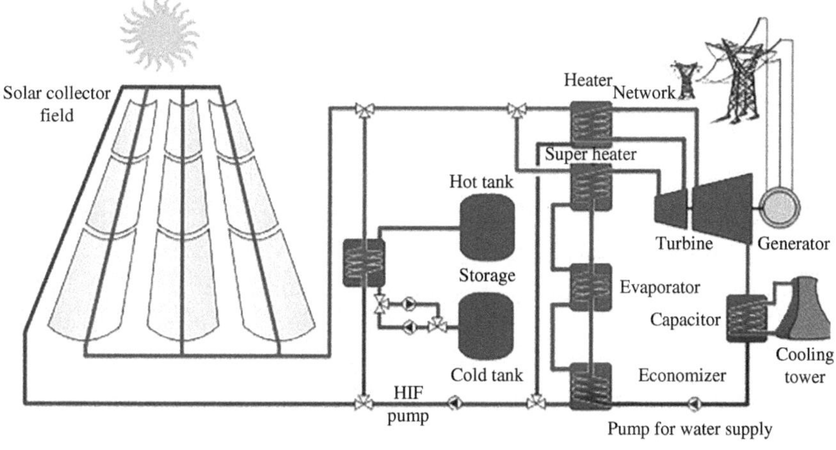

Fig. 7.2 power plants with a parabolic reflectors channel.

obtained in this way is not suitable for transmission over long distances, nor is it enough to run high-power devices (Jakhongir & Ugli, 2019).

Application of PV systems

PV systems are very diverse; they can be smaller than a coin, larger than a football field, and can provide power for any device, from clocks to entire settlements. With the ease of handling, these factors make them particularly attractive for a wide range of applications.

The recent increase in production of low-cost PV cells has opened many new markets with a large number of different applications. Applications such as lighting, telecommunications, cooling, pumping water, and providing electricity for a whole settlement, especially in remote areas, have proven to be competitive and profitable in comparison to existing technology. Additionally, a relatively new application of these systems with very high potential—PV façade systems (BIPV—building integrated PVs)—has appeared.

Facade PV systems

The more current aspect of electricity generation is the conservation of the environment in which energy is generated and consumed. Solar electricity can contribute to available energy and at the same time help in preventing global climate change. Approximately 75% of energy used in the developed world is spent in cities, of which about 40% is consumed in buildings. PV systems can be embedded in almost any building or structure, from bus shelters to large office buildings and even gardens, parks, and so on. Although the exact forecast of PV performance in buildings requires a careful analysis of various factors such as quantity of solar radiation coming to the surface of the building, stability and quality of electrical installation, electrical and distribution networks, etc., it is easy to see that this technology has great potential. Even in climatic conditions characterized by moderate sun radiation, the roof of a building of a household is large enough to set a PV system that can provide enough electricity during the year. PV modules and generators are traditionally placed on special supporting structures, but they can also be placed on buildings, or they can become integral parts of buildings (Salpkari & Lund, 2016). The use of PV systems can significantly reduce the consumption of electrical energy from power plants. The buildings can even be converted into small manufacturers and distributors of electrical energy that can be of general benefit.

The problem of architectural integration of PV technology requires an interdisciplinary approach. This requires not only collaboration and the presence of highly specialized experts on the project team, but also consideration of sensitive issues related to their social, economic, and energy aspects. For example, the facade of the building not only has to protect against precipitation and regulate heat losses, but it must also regulate the flow of sunlight, provide sound insulation, provide ease of maintenance, and must also meet architectural and aesthetic criteria.

As facade PV modules can replace traditional building materials, the difference in cost between the solar elements per unit area and materials that can be replaced is of particular importance. As the cost per unit area of a facade PV system, connected to the distribution network, is almost the same as the price of the highest quality materials, such as marble or ornamental stone, the additional benefits of PV are practically free (Wang et al., 2017).

Potential of PV devices

Solar PV technology was previously used mainly in space programs or in remote locations, and thus was marginal and exotic at first. However, in the last 10 years, it has become the main technology for the production and distribution of electric power in urban areas with the potential to become equally cost competitive with the prices of energy obtained and distributed by conventional technologies.

An increasing number of companies and organizations are actively involved in the promotion, development, and manufacture of PV systems. Companies that produce and distribute electric power in cooperation with manufacturers of solar equipment, urban organizations, and supporting national and international funds are planning and implementing increasingly large projects. The market value of the PV industry currently amounts to over $5 billion a year, and it is expected to increase to over $10 billion per year by 2010. Representatives of the PV solar industry are some of the leading global companies, such as Sharp, Mitsubishi, Sanyo, BP, Shell, Kyocera, etc.

The market is dominated by urban (residential) PV systems connected to the electricity network. In the period from 2000 to 2005, there was a relative increase of 50% per year in the number of networked solar PV power plants. Moreover, studies of the European Photovoltaic Industry Association (EPIA) and the Greenpeace organization predict that half the capacity of 207 GW in 2020 will be systems that are connected to the electrical distribution network, out of which 80% will be installed in residential buildings. The PV industry is increasingly present in national energy strategies of more and more countries.

Biomass energy

Energy plays a crucial role in modern life. It is needed for heating, lighting, and cooking in households and for virtually every industrial, commercial, and transport activity. At the global level, consumption of energy grew steadily by around 2% a year in the decade 1990–2000 and continues to grow. Fossil fuels (coal, gas, and oil) currently account for about 79% of world energy consumption, nuclear energy for 7%, and renewable energy sources for 14%. Global consumption of fossil fuels grew in line with overall energy consumption during the 1990s and is expected to grow faster than overall consumption in the period up to 2020. Fossil fuels have two main disadvantages. First, when they are burned, they emit pollutants, including the greenhouse gases (GHGs) that are causing climate change. Second, countries without adequate reserves of fossil fuels are facing increasing risks to the security of their energy supplies. Using renewable energy sources in place of fossil fuels reduces emission of GHGs and other pollutants, improves security of supply by boosting diversification of energy production, and encourages the creation of new jobs and businesses. In the 25 countries of the EU, renewable energy sources provided about 6% of total energy requirements in 2002. Currently, nearly two-thirds of all energy from renewable sources used in Europe comes from biomass, and this source is set to play a significant role in meeting the 2010 target (Report and Data Portal, 2021).

What is biomass?

Biomass covers a wide range of products, by-products, and waste streams from forestry and agriculture (including animal husbandry) as well as municipal and industrial waste streams. A definition adopted by EU legislation is: "…the biodegradable fraction of products, waste and residues from agriculture (including vegetal and animal substances), forestry and related industries" Biomass thus includes trees, arable crops, algae, other plants, agricultural and forest residues, effluents, sewage sludge, manures, industrial by-products, and the organic fraction of municipal solid waste (MSW) (García, Pizzaro, Lavín, & Bueno, 2017).

Photosynthesis and the carbon cycle

All life on earth is based on green plants that convert carbon dioxide (CO_2) and water from the atmosphere into organic matter and oxygen, using the energy of the sunlight. This process is called photosynthesis. Expressed simply, the overall chemical reaction is as follows (in this example glucose is the product):

$$6CO_2 + 6H_2O = C_6H_{12}O_6 + 6O_2$$

In plants, the energy from the sun is conserved in the form of chemical bonds. This chemically stored energy can be used by the plants themselves or by animals and human beings.

When biomass is burned or digested, the organic carbon is recycled in a complex global process known as the carbon cycle. In this process, the CO_2 that was absorbed as the plants grew is simply returned to the atmosphere when the biomass is burned. Therefore, if the cycle of growth and harvest is maintained, there is, broadly speaking, no net release of CO_2 (Singh, Lakhi, et al., 2019; Singh, Tingas, Goussis, & Im, 2019). This is why biomass is regarded as an energy source that does not emit CO_2 into the atmosphere when burned (see Fig. 7.3).

Fossil fuels, of course, are also organic matter. However, in their case the matter has been transformed and stored over a long period of time under heat and pressure in the absence of oxygen. When we burn fossil fuels, we release in a short period a quantity of CO_2 that has been locked up in plants and their follow-up products over millions of years.

Biomass has many advantages as an energy resource. It can be used to produce a wide variety of product types—heat, electricity, solid fuels, liquid transport fuels, gaseous fuels, and other products. Biomass raw materials come in a range of forms that are abundant in most parts of the world, including Europe. Their use does not increase the CO_2 content of the atmosphere. As a result of work done over the last decade, some excellent modern processes now exist for converting raw materials to usable products such as the biodiesel and bioethanol blends that can be used in today's vehicles without engine adaptation. In addition, some high-technology equipment has been developed especially for biomass product use—for example, fully automated boilers suitable for burning wood pellets (Geng et al., 2019). A research framework is in place that will allow further technological progress to be made. Some legislative measures to support the use of bioenergy, i.e., energy derived from biomass, have already been introduced. Other policy steps needed to allow widespread use of bioenergy products are being developed.

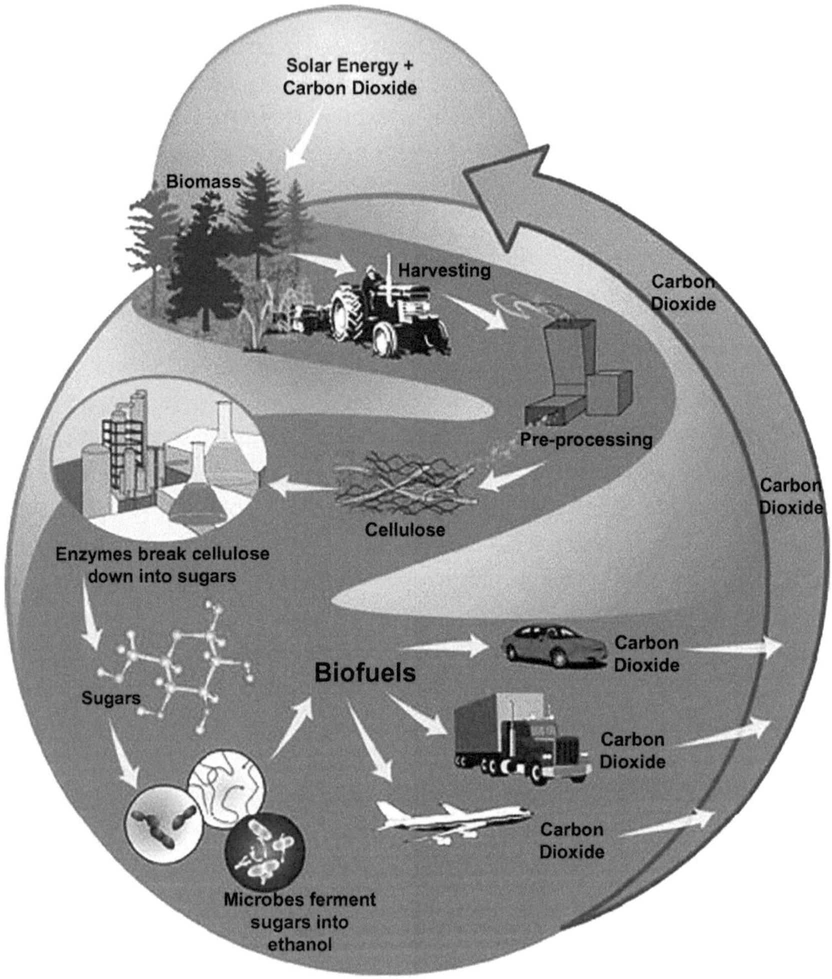

Fig. 7.3 Life cycle analysis of CO_2.

Of course, not all the existing biomass resource can be used for energy purposes. Food, timber, paper, and board and certain high-value chemicals are also derived from biomass. Therefore, bioenergy production must be integrated with other priority applications. Biomass must be used in a wise and sustainable way.

The overall aim is to show where and how bioenergy products can be used right now, the high technology nature of today's processes and equipment, the role of research in developing these, the benefits of using biomass for energy purposes, and what needs to be done from the policy and research viewpoint to ensure full development of this sector in the future. Following this introduction, the section is arranged in three main subsections covering biomass resources, conversion processes, and products.

Biomass resources

Biomass comes in various forms, each of which has specific properties, uses, and advantages. The main sources are wood from conventional and short-rotation forestry, other energy crops, residues from forestry and agricultural production, and by-products and wastes from industrial and municipal processes.

Availability and potential

Total EU land area is around 385 million acres. Forests and woodlands cover 137 million acres and crops 178.5 million acres. Once the requirements of the food, wood products, and paper sectors have been met, the biomass resources from these trees and crops could provide around 8 EJ (1 exajoule = 1.0E + 18 joules) energy a year—about 11% total annual EU energy consumption. In practice, we are exploiting less than a quarter of available resources.

If biomass is to play its expected role in achieving the EU target for renewable energy in future, much more will need to be done both to exploit the existing bioenergy resources and to establish new ones. Integrated production of wood and biomass could be introduced where the trees are thinned to maximize the value of the wood produced and the thinning used for biomass. Increased areas of land could be used for cultivating energy crops, i.e., short rotation forestry or other crops. Finding the land for growing energy crops is an important issue. Various woody and nonwoody plant species are available that are suitable for different qualities of land. In recent years, surplus food production in the EU has led to cropland being left fallow. In 2001, for instance, as much as 15% (i.e., 5.7 million acres) of the EU-15 (The EU15 comprised the following 15 countries: Austria, Belgium, Denmark, Finland, France, Germany, Greece, Ireland, Italy, Luxembourg, Netherlands, Portugal, Spain, Sweden, United Kingdom) cropland was under voluntary set aside. Other options may exist, such as growing nonfood crops. Recent reforms of the EU's agricultural policy encourage production of energy crops. An extra subvention of 45 Euros per acre can now be given for energy crops for a guaranteed area of 1.5 million acre (Hamelin, Borzęcka, Kozak, & Pudetko, 2019).

Wood

Wood is a renewable raw material that can be used not only for making timber-based products, pulp, and paper but also as a source of energy. In traditional professionally managed forests, the normal life cycle of the tree includes plantation, a stage of rapid height growth followed by a stage of steady growth in diameter, height, and volume. The point of harvesting depends on the species, but it is generally reached after 30–80 years. Twenty-five percent to forty-five percent of the wood harvested each year is in the form of residues, i.e., the wood from forestry thinning and the residues from felling. Therefore, most of the woody biomass resources available from forests and woodlands for energy production consist of residues. In the EU, woody residues are estimated to have the potential to provide 3.8 EJ energy annually (Sulaiman, Abdul-Rahim, & Ofozor, 2020).

For certain species, the use of short-rotation techniques can reduce the life cycle of the tree to 3–15 years. Thus, plantations dedicated to short-rotation forestry can provide an important additional source of woody biomass for energy purposes. The species most commonly used in this way include poplar, *Salix*, willow, and eucalyptus.

The key to creating an economic energy-from-woody-biomass scheme is to establish effective logistical systems for harvesting, recovering, compacting, transporting, upgrading, and storing the wood. Harvesting and transport, in particular, can have significant impacts on energy balance and costs. The trend is to move toward greater mechanization of harvesting for reasons of economy and safety. Because firewood and forest residues are low-value commodities, transport costs constitute the most important part of total production costs. Care must be taken, therefore, to choose an appropriate method of transport and locate the conversion plant as near as possible to the woody biomass source (González-García & Bacceneti, 2019).

Nonwoody energy crops

Nonwoody plants most often used for energy purposes are wheat, barley, rye, sugarcane, sugar beet, leguminous plants (e.g., alfalfa or lucerne), grass (e.g., miscanthus and switch grass), and oil crops (e.g., rapeseed). Many other species have been studied with respect to agronomy, yield optimization, harvesting, storage, and processing. These include high-yield fibrous plants such as giant reed and a form of globe artichoke.

These plants can provide biomass suitable for direct combustion or thermochemical or biological conversion. Wheat, barley, rye, sugarcane, and sugar beet, for instance, are generally converted to ethanol. Leguminous plants and grasses can be processed together with manure or waste to obtain biogas for heat, electricity, or fuel. Oil crops can be used to produce biodiesel.

There are also plants that can be processed to give liquid biofuels and cellulosic materials simultaneously. An example is sweet sorghum, which yields both bioethanol and dry cellulosic material for other bioenergy use. Some of the plants listed above are perennials. Others are annuals. All can fit into conventional agricultural practice. Cultivating crops for energy use does not preclude a farmer from growing food crops as well—or vice versa. Food and energy crops can usefully be grown hand in hand to maximize the overall efficiency of the farm. The advantages of energy crops over food crops are that they do not require the best land and need significantly less care, water, and fertilizers. This is because it is the quantity, rather than the quality, of the product that is important (Pandey, Bajpai, & Singh, 2016).

Agricultural residues and by-products

As indicated above, the residues from forestry thinning and felling constitute a major biomass resource. To this must be added the different types of residue and by-products that come from processing wood, e.g., sawdust, bark chippings, wood shavings, plywood residues, and black liquor. (The latter, a by-product of paper manufacturing, can be burned or gasified to produce fuels for transport.)

Residues from the harvesting and processing of food and other crops—cereals, sugarcane, tea, coffee, rice, cotton, rubber trees, and coconut palms—are also important. It should be noted, however, that only about 20% of total straw can be used for energy purposes, to ensure that the demands of the agricultural sector and other markets can be fulfilled. Animal manure is another useful resource from the farming sector.

Together, these resources have the potential to provide the EU with 6.75 EJ energy a year—56% from wood residues and by-products, 8% from straw, 25% from other crop residues, and the rest from animal manure (Bedoić, Ćosić, & Duić, 2019).

Marine biomass

Plankton, algae, and other marine-based organisms constitute a biomass resource that has not yet been exploited. This area is, however, the subject of continued research. Bearing in mind the volume of the sea, this resource could provide a major carbon-neutral source of energy for the future.

Wastes

MSW is primarily waste produced by domestic households, although it also includes some commercial and industrial wastes that are similar in nature. Each EU citizen produces on average of more than 500 kg of MSW a year. The total MSW resource in the EU is therefore of the order of 225 million tons a year (Scarlat, Fahl, & Dallemand, 2019).

The organic fraction of MSW has a significant heat value. Typically, MSW has a heat value of 8–12 MJ/kg, about a third of the heat value of coal. A ton of MSW will give about 2 GJ of electricity.

Decisions on whether to use MSW as an energy resource are linked to local and national waste management policies and the views of the public on, say, recycling and incineration.

The choice of MSW treatment for a particular locality must take into account, among other things, the composition and properties of the input waste, the available technologies, and the market for the various outputs. The whole process must be integrated to avoid conflicting claims on the different streams (see Fig. 7.4).

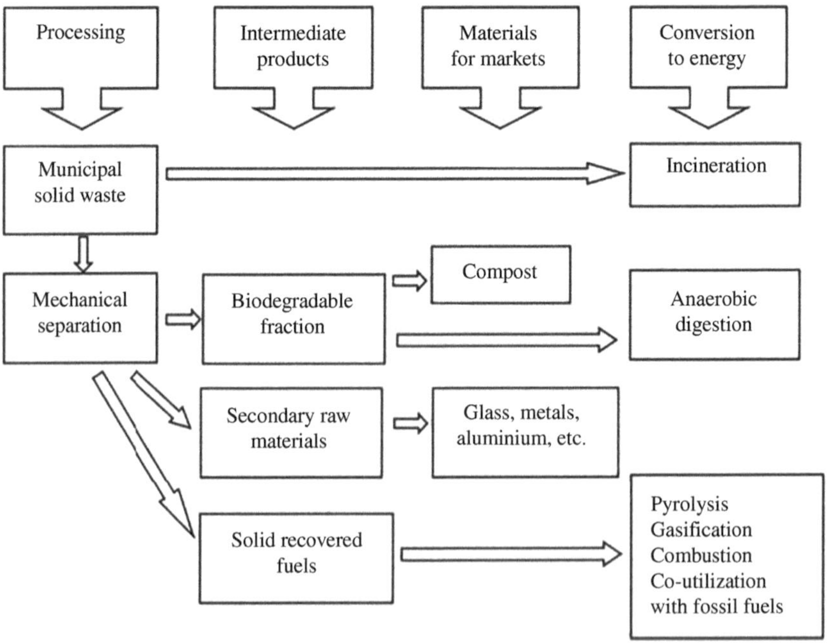

Fig. 7.4 Pathways for MSW treatment and conversion.

In general, for wastes with a high energy content, the major part will be used for power and heat production. Incineration offers one route for doing this. In 2002, there were approximately 340 incineration plants in the EU handling between them 50 million tons of waste a year. Recent installations tend to be efficient in generating power or producing combined heat and power. Another option is to convert the waste to solid, liquid, or gaseous fuels that can be more easily transported and used to produce heat or power or to drive vehicles. The biodegradable part of MSW can be used to produce compost or digested along with other suitable wastes to give biogas. Biogas is another useful source of energy. It can be recovered from landfill sites or produced by digesting not only parts of MSW but also sludge from sewage treatment, livestock manure, and suitable agricultural and agro-industrial effluents. It has been estimated that the total energy content of all the above resources capable of producing biogas in the EU exceeds 3.35 EJ. It should be possible to mobilize 0.63 EJ by 2010 and 0.75 EJ a year by 2020, with expectation of growth. More widespread exploitation of biogas would also be in line with European environmental policy since it would reduce emission of methane, a powerful GHG, to the atmosphere (Araci, Haro, Giuntoli, & Ollero, 2017).

Biomass conversion

Except for cases where straightforward combustion is appropriate, it is not usually possible to use biomass raw materials as they stand. They have to be converted in some way to solid, liquid, or gaseous fuels that can be used to provide heat, generate electricity, or drive vehicles. This conversion is generally achieved by some type of mechanical, thermal, or biological process.

Mechanical processes are not strictly conversion processes since they do not change the nature of the biomass. They are commonly used in the treatment of woody biomass and waste. The sorting and compaction of waste, the processing of woody residues into bundles, pellets, and chips, the cutting of straw and hay into pieces, and the squeezing of oil out of plants in a press are all examples of mechanical processes. Such processes are often used to pretreat a biomass resource for further conversion. They are therefore discussed, where relevant, later in this section in conjunction with other methods of conversion or when describing the final products.

Combustion, gasification, and pyrolysis are examples of thermal processes. They produce either direct heat or a gas or oil, such as bio-oil. The gas can be used to drive a motor or a fuel cell. The bio-oil can be further transformed into gaseous and liquid fuels.

Fermentation and digestion are examples of biological processes. These use microbial or enzymatic activity to convert sugar into ethanol or biomass to solid fuels or biogas. The following sections highlight some major thermal and biological conversion processes (see Fig. 7.5) (Mossier & Ileleji, 2020).

Thermal processes

Processes where biomass conversion is achieved by heat are the most commonly used technologies.

Combustion is the most ancient and frequently applied way of using biomass as an energy source because of its low cost, ease of handling, and high reliability. The biomass can either be fired directly (as when firewood is burned for heating or waste is incinerated) or cofired with fossil fuels. Modern coal-fired plants are increasingly being designed for cofiring in order to reduce carbon dioxide emissions (Zaporozhets, 2020).

Factors to be considered when designing a biomass combustion system include the characteristics of the fuel to be used, environmental legislation, the cost and performance of available equipment, and the output required. During combustion, a biomass particle first loses its moisture at temperatures up to 100°C using heat given off by other particles. Then, as the dried particle heats up, volatile gases containing hydrocarbons, carbon monoxide (CO), methane (CH_4), and other gaseous components are released. In the combustion process, these gases contribute about 70% of the heating value of biomass. Finally, the char oxidizes, leaving ash. The technical aspects of

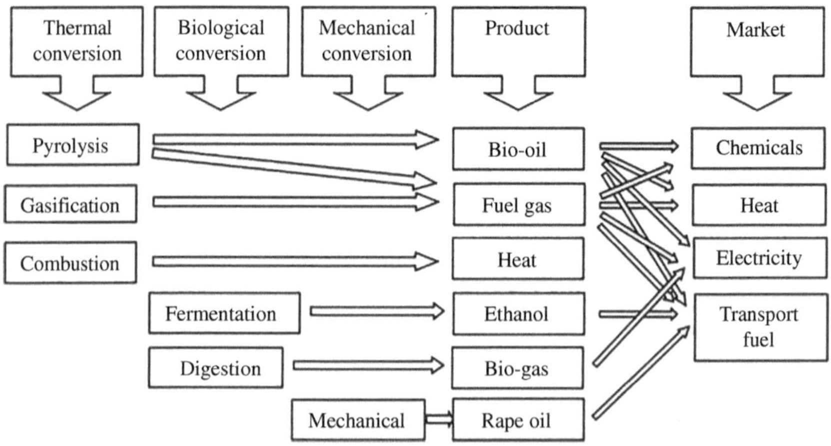

Fig. 7.5 Biomass conversion pathways.

heating and large-scale combustion are discussed in more detail in the section on heat and power production from biomass, as follows (see Fig. 7.6).

The basic thermochemical processes are gasification and pyrolysis. Both processes involve heating the feedstock in the presence of less oxygen than is required for complete combustion and produce a mixture of gas, liquid, and char. Yields of the various outputs depend on the nature of the biomass used, the rate of heating, the highest temperature reached, the way in which off-gases react with hot solids, the amount of water (as steam), and the presence or absence of other substances that may act as catalysts. At one extreme, processes can be optimized to produce charcoal. At the other, they can be designed to produce a mixture of hydrogen and carbon monoxide (synthesis gas) suitable for use in the catalytic formation of a variety of liquid fuels (Wang, Jian, Shu, Chen, & Chen, 2020; Wang, Zhang, Mao, & Gong, 2020).

In pyrolysis, an external source of heat is used with no oxygen. Heat causes the biomass molecules to break down to form water (steam) and highly reactive low-molecular-weight fragments, many of which combine to form char. Hot char will react with steam to produce carbon monoxide and hydrogen, giving a gas with a high heating value. Some of the low-molecular-weight compounds may be swept from the reactor, recombining to form tars as they cool. Fine particles of ash and partly carbonized biomass may also be carried in the gas stream (Shurong, Gongxin, Haiping, & Zhongyang, 2017). For these reasons, the gas requires advanced cleaning before it can be used in a combustion engine or turbine (see Fig. 7.7).

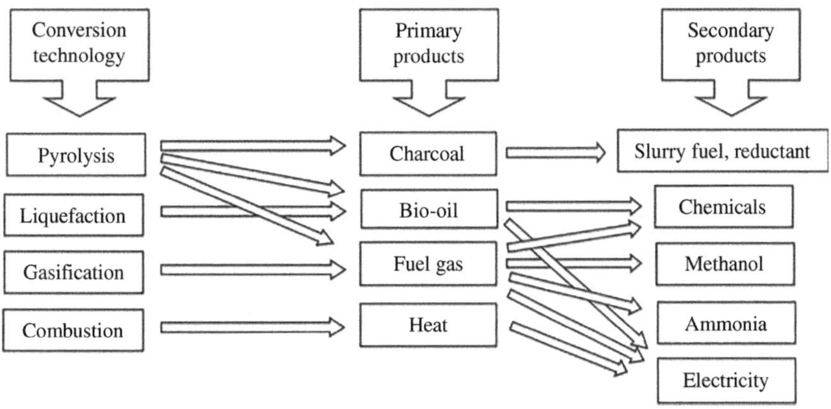

Fig. 7.6 Thermochemical conversion products and uses.

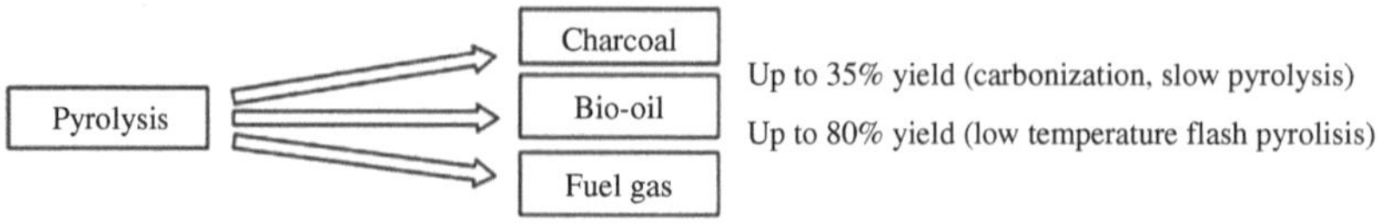

Fig. 7.7 Pyrolysis products.

Fast pyrolysis is a high-temperature process in which small particles of biomass are burned rapidly in the absence of oxygen causing them to decompose to give vapors, aerosols, and some char. After cooling and condensation, a dark viscous liquid (bio-oil) is obtained. It has a heating value about half that of conventional fuel oil and can substitute for the latter in combustion systems or engines for heat or power generation. Further processing of the oil by hydrogenation or using a catalyst will give a higher-quality product close in specification to petroleum-derived fuels oils. This product can be used in diesel-fired vehicle engines.

Liquefaction is a low-temperature, high-pressure thermochemical conversion process carried out in the liquid phase, which has the potential to produce high-quality products. It requires the use of either a catalyst or hydrogen under high pressure. Gasification is a high-temperature thermochemical process carried out under conditions that lead to a combustible gas, rather than heat or a liquid. Modern gasifiers can use a large variety of feedstocks. The process involves partial combustion of the biomass feedstock with a restricted supply of air or oxygen at temperatures in the range 1200–1400°C consisting mainly of carbon monoxide, hydrogen, methane, and carbon dioxide. However, most biomass gasifiers use air for cost reasons so the output is diluted with nitrogen and therefore has a lower heating value (Gollakota, 2018). Whichever process is used, the product can, after appropriate treatment, be burned directly or used in gas turbines or engines to produce electricity or mechanical work. The process can be varied to give a hydrogen-rich gas or synthesis gas that can be used to make other fuel products.

Gas cleaning

This is a critical step in both combustion and gasification systems. The aim is to reduce emissions in flue gases, reduce the level of damaging contaminants (e.g., hydrogen sulfide and mercaptans) in biogas and landfill gas, and remove particles and tars from gas generated by chemical processes. A wide range of techniques are available, including gas scrubbing with water and chemical solutions; filtration; electrostatic precipitation or use of cyclones to remove particles; use of molecular sieves or chilling to remove water and other impurities; and use of iron, calcium, or zinc oxide or

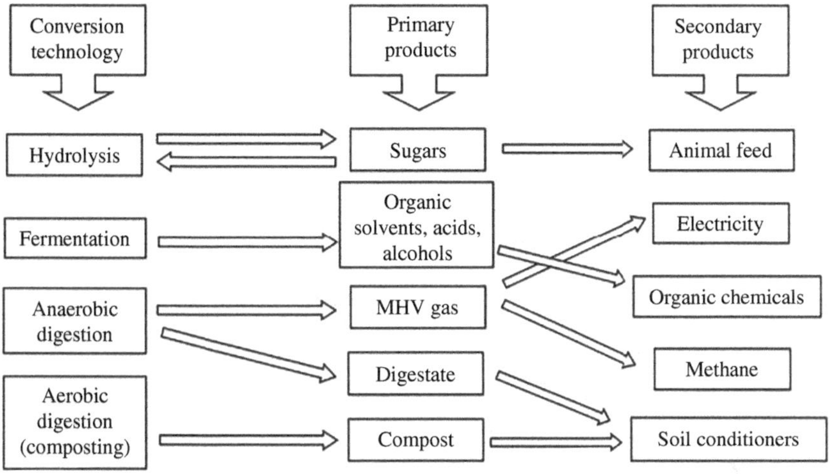

Fig. 7.8 Biochemical conversion product and uses.

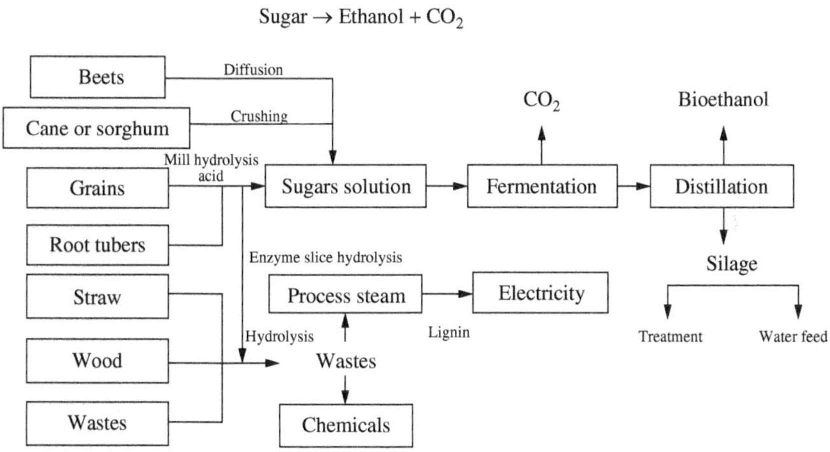

Fig. 7.9 Fermentation.

chemical reduction to remove sulfur compounds. In particular, tars may be cracked by passing the gas stream back through the gasifier bed, or through a second stage, with external heating. Scrubbing with water or various proprietary liquids may also be used to reduce levels of carbon dioxide. Most of these processes are commercially viable when used on a large scale, as they are in the petrochemical industry (Lotvi, Ma, Austin, & Kumar, 2019). They may not, however, be economically feasible when applied to small biomass-based facilities (see Figs. 7.8 and 7.9).

Nonthermal processes

A number of processes are available that use some type of biochemical process, rather than heat, to achieve biomass conversion.

In anaerobic digestion, a process that takes place in the absence of oxygen, a mixed population of bacteria catalyzes the breakdown of the polymers found in biomass to give biogas (MHV gas). This primarily consists of methane and carbon dioxide but may also contain ammonia, hydrogen sulfide, and mercaptans, which are corrosive, poisonous, and odorous. The process takes place in several stages. First, polymers such as cellulose, starch, proteins, and lipids are hydrolyzed into sugars, amino acids, fatty acids, etc. These are then converted to a mixture of hydrogen gas, low-molecular-weight acids (primarily acetic acid), and carbon dioxide, in the process of acetogenesis. Finally, these products react together to generate methane, in methanogenesis (Paul & Dutta, 2018). Millions of small anaerobic digesters have been built in rural areas worldwide with methane generation as their primary aim.

Fermentation is one of the oldest biological processes used by mankind. It normally uses yeast (an organism that secretes catalytic enzymes) to initiate chemical reactions that lead to the desired outputs—ethanol and carbon dioxide. Ethanol is used not only for alcoholic drinks but also as a solvent, additive, and fuel. Bioethanol as a transport fuel is discussed in more detail later in this chapter in the section on liquid biofuels. The main producers are Brazil (which uses sugarcane as feedstock) and the United States (which uses corn). Some fuel alcohol is produced in Europe using wheat, molasses, and petrochemical feedstocks.

Breakdown of the biomass into useful products can be speeded up by the use of enzymes other than yeast. These enzymatic conversion processes are mainly used for degradation of starch, cellulose, proteins, or lipids feedstocks.

The biorefinery

A biorefinery is defined as a facility for achieving large-scale integrated production of fuels, power, and chemicals from biomass. It is analogous to a petroleum refinery, which produces multiple fuels and products from petroleum. The possibility of integrated production of a number of products from biomass is a concept that is gaining increased attention in many parts of the world. In the EU, the developments toward a carbon-constrained economy and evolving agricultural policy make the idea particularly interesting. No large-scale biorefineries exist at present, but they are regarded as being an important element in the future of biomass. Work has been carried out in the EU, the United States, and elsewhere on the design and feasibility of such facilities. The concept is explored in the following. A biorefinery is based on a number of conversion processes (Infelise, Kazimierczak, Wietecha, & Kopania, 2019).

The biorefinery requires year-round supply of biomass feedstock, preferably of a specific quality. Possible sources of feedstock include agricultural crops and residues, wood residues, and woody and nonwoody energy crops. If supplies are of heterogeneous type and quality, the feedstock has to be mechanically separated into fractions and, where necessary, pretreated to give interim products of the required specification for the different applications. The final conversion to energy, fuels, or other products is carried out using a range of thermochemical and biochemical processes—some of which are already at a stage of commercial development, while others require further research and technological development.

A range of products are delivered with multiple ends uses, including low-volume and high-value specialty chemicals that have niche uses in the food and other industries; high-volume and low-value liquid fuels for widespread use in the transport industry; heat; electricity, etc. The diversity of the products gives a high degree of flexibility to changing market demands and allows plant operators many options for gaining revenues and achieving profitability.

In addition, there are economies of scale. Advanced conversion of biomass (gasification, pyrolysis, etc.) has proved costly to date. The large-scale operations conducted in the biorefinery offer cost savings by, for example, allowing preconversion feedstock treatment facilities to be shared. A large-scale operation has greater buying power when acquiring feedstocks: it can source biomass over a larger geographical area and negotiate cheap long-term contracts. In the immediate future, biorefinery products will not, as a rule, compete in cost terms with products made from fossil fuels. Rather, their main competitive advantage comes from environmental sustainability. In particular, biorefinery products are near-neutral in terms of GHG emissions. Assessing the financial advantage this will bring in 10 or more years' time is very difficult (Longati, Lino, Giordano, Furlan, & Cruz, 2018).

For biorefineries to succeed, different sectors of the economy—agriculture, forestry, agro- and wood-based industries, chemical, food, transport, and energy industries—will need to cooperate to develop processes for making new biomass-based products and bring them to market. There is a need for extensive research and technological development to test and prove the supply of biomass feedstocks, the extensive range of biorefining technologies, the end uses of the products, etc. Research institutes and universities are therefore also important stakeholders. Policymakers, regulators, and lawmakers will also play an important enabling role in establishing the biorefinery concept. As discussed later in this chapter, today's hydrocarbon-based economy could well evolve into a bio-based economy where biorefineries play a very important role (see Fig. 7.10).

Products from biomass

As explained above, biomass can be converted by a variety of processes to a wide range of products—including heat, electricity, and solid, liquid, and gaseous biofuels. The following sections give more information on these products and the ways in which they can be made, supplied, and used, and the current status of their application in the EU. Solid and gaseous biofuels are discussed first because their main applications are as intermediates in the preparation of other products, i.e., heat, electricity, and liquid biofuels. At the end of this chapter, a short discussion on biomass products that are not energy-related is given (Pandey et al., 2019).

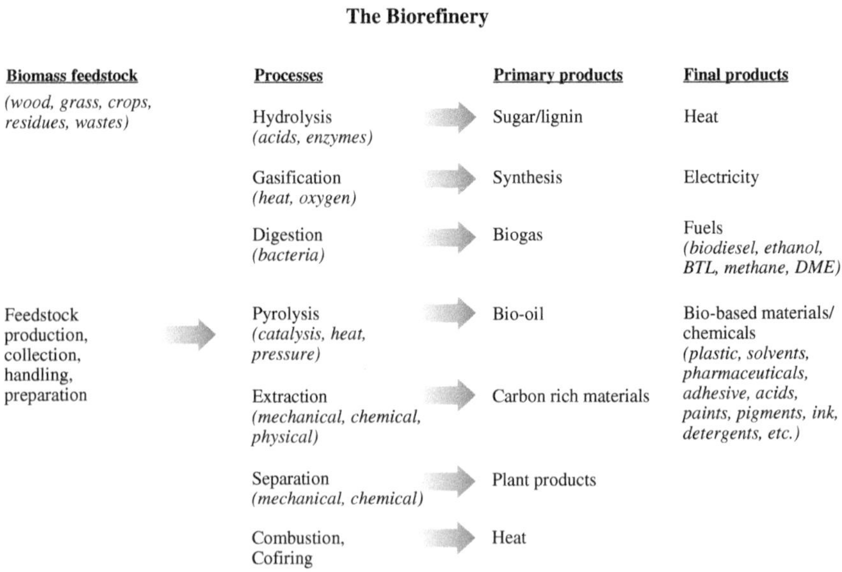

Fig. 7.10 The biorefinery.

Solid biofuels

Solid biofuels can be derived from many biomass resources such as wood and wood residues and by-products, agricultural crops and residues, and the combustible part of solid wastes.

The quality of solid biomass as a fuel is related to properties such as moisture content, heating value, ash content, and content of volatile matter. The higher the moisture content of a fuel, the lower its heating value. This is partly because fuels with a high moisture content have, by definition, a lower content of combustible material. It is also because some of the heat liberated during combustion is used up in evaporating the water. For a fuel to be capable of being ignited and having energy extracted from it, its moisture content must be below 55%. The moisture content of biomass sources ranges from less than 10% for straw to 70% for forest residues and wet waste. The moisture content of bark and sawmill products is in the range of 25%–55%. By contrast, the moisture content of processed wood pellets is less than 10% (Motta, Miranda, Filho, & Maciel, 2018).

Ash content is significant because it determines the behavior of the biomass at high temperatures: quantities of molten ash can, for instance, cause problems during combustion. The ash contents of biomass can range from 0.5% for wood, through 5%–10% for energy crops, to 30%–40% for agricultural residues such as husks (Liu, 2019).

When subjected to heat, biomass decomposes into volatile gases and char. The volatile matter content of a resource is described by the proportion that volatilizes at 400–500°C. Typically, biomass has a volatile matter content of over 80%, compared with 20% for coal.

The formation of nitrogen oxides (NOx) during combustion can be a particular issue with biomass fuels. When any fuel is subjected to combustion in the presence of air, some NOx will be formed as a result of the reaction of the nitrogen in the air with the fuel at temperatures above 950°C. In the case of biomass fuels, however, some NOx will also be formed at low temperatures because of the presence of nitrogen in the biomass itself. The quantity of NOx formed can generally be controlled by using appropriate combustion techniques.

Wood is the most commonly used solid biofuel. The raw material can be in the following forms: logs, stems, needles, and leaves from the forest; bark, sawdust, and redundant cuttings from the sawmill; chips and slabs from the wood industry; and recycled wood from demolition. These can be used directly as a fuel, where this is appropriate. Alternatively, they can be processed into forms that allow for easy transport, storage, and combustion, such as chips, pellets, briquettes, and powder. Firewood is forest fuel in the form of treated or untreated stem wood. A new technique that allows for easier handling is bundling, where branches are pressed together into log-like bundles of equal size. Wood powder consists of ground wood raw material. Wood chips are pieces of wood about 1–5 cm in diameter. Pellets are short cylindrical or spherical pieces with a diameter less than 25 mm. Pellets are produced from sawdust, cutter shavings, chips, or bark by grinding the raw material to a fine powder that is pressed through a perforated matrix. The friction of the process provides enough heat to soften the lignin present. During the subsequent cooling, the lignin stiffens and binds the material together. Wood briquettes are rectangular or round pieces made by pressing together in a piston press finely ground sawdust, cutter shavings, chips, bark, etc. (Danish & Ahmad, 2018).

The energy content of pellets and briquettes is around 17 GJ/ton with a moisture content of 10% and a density of around 600–700 kg/m^3. To replace oil, one needs about three times its volume in pellets (Miranda et al., 2015).

Using wood-derived biofuels instead of fossil fuels is helpful from the viewpoint of aiding sustainability, reducing GHG emissions, and improving quality in forestry. Since the major users of solid biofuels are companies concerned with heat and electricity production, better integration of the forestry industry with energy companies would undoubtedly improve the overall efficiency of the scheme. If, for instance, bulky materials with a high moisture content have to be transported far, the costs of using solid biofuels are increased considerably.

It is possible to use many agricultural crops and residues as solid fuels. Examples are straw, husks, stalks, bagasse from sugarcane, and grass. Using these residues as fuels can in addition solve the problem of how to dispose of them.

As described earlier, the initial sorting of municipal solid waste results in the recovery of combustible solids that can be used as fuels.

Gaseous biofuels

The most commonly used biofuels are biogas and hydrogen. Both are produced from biomass wastes by biochemical processes (see Fig. 7.11).

Biogas

Chemically, biogas comprises a mixture of hydrocarbons (mainly methane) and other gases. It can be produced by anaerobic digestion of sewage sludge, grass, and other ley crops, manure, and agricultural and food wastes, including

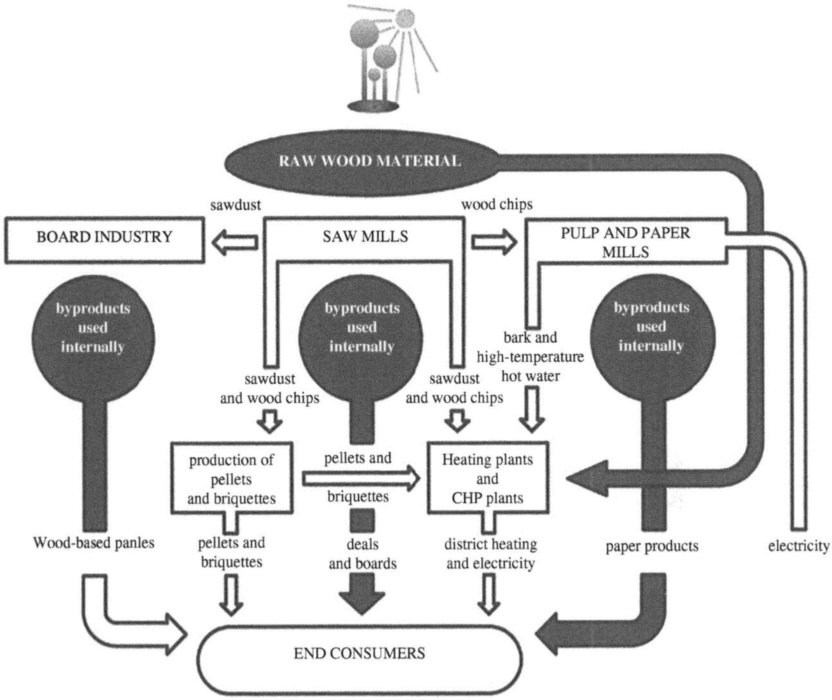

Fig. 7.11 The flow of products, by-products, and energy within the forest industry.

those from slaughterhouses, restaurants, grocery stores, and wastes from the pharmaceutical industry. It can also be extracted from landfills, where it is formed spontaneously and, if left, would cause environmental problems since methane is a powerful GHG (Nevzorova & Kutcherov, 2019).

Biogas production plants based on agricultural wastes can be found in the countryside in most EU member states. Their unique smell has given rise to the unfortunate and untrue notion that biomass is not a clean source of energy. The availability of landfill as a source of biogas will, of course, vary from member state to member state according to national policies on the use of landfill as a means of disposing of organic wastes. In some countries, landfill gas is recovered with fully industrial technologies. Frequently, biogas is used close to the place where it is produced. Its main applications are for production of heat, electricity, and combined heat and power. Further information on these is given later in this chapter. Biogas's main advantage over other biomass-derived fuels is that it can be burned directly in any gas-fired plant. It can also be injected into the natural gas network. In addition, biogas can be used as a transport fuel for vehicles adapted to run on gas. The environmental benefits of replacing petrol or diesel with biogas are considerable. However, although the cost of biogas is significantly lower than that of petrol per unit of energy, one has to equip vehicles with extra gas tanks to use it.

Biohydrogen

The EU devotes part of its energy research budget to exploring the possibilities for hydrogen in the community's future energy supplies. Its potential is thought to be considerable. Hydrogen is a clean fuel that does not emit any GHGs during combustion: the remnant of combustion is pure water. Hydrogen can be burned directly to produce heat and electricity. It can be used as a transport fuel. It also makes an ideal input for fuel cells, where it is converted directly to heat and power with high efficiency. Further information is given in the electricity section later in this chapter.

Biomass is potentially an important source of renewable hydrogen. Hydrogen can be produced from a broad range of biomass sources containing carbohydrates, cellulose, or proteins using biological processes. Under anaerobic conditions (i.e., when no air is present), bacteria are able to convert the biomass to hydrogen, biogas, and ethanol. Typical yields are in the range of 0.6–3.3 molecules hydrogen per molecule of glucose, depending on the particular bacteria used. Thermophilic bacteria that operate at temperatures up to 70°C give higher yields than bacteria that operate at ambient temperatures (Davood, Anda, & Shafiullah, 2020).

A typical chemical reaction is:

$$C_2H_{12}O_6 + 2H_2O = 2CO_2 + 2CH_3COOH + 4H_2$$

The yields can be increased further by using phototropic bacteria that convert acetic acid to hydrogen, as follows:

$$CH_3COOH + 4H_2O = 2CO_2 + 4H_2$$

As a rule, biological processes require modest investments and are effective even on a small scale (see Fig. 7.12).

Heat

Wood-based solid biofuels are the biomass sources most commonly used in heat production. Combustion is the normal conversion method. Traditionally, of course, firewood was used for thousands of years to provide heat for domestic purposes, i.e., local heating and food preparation. Today, the availability of biomass-derived fuels in clean and convenient forms (e.g., chips,

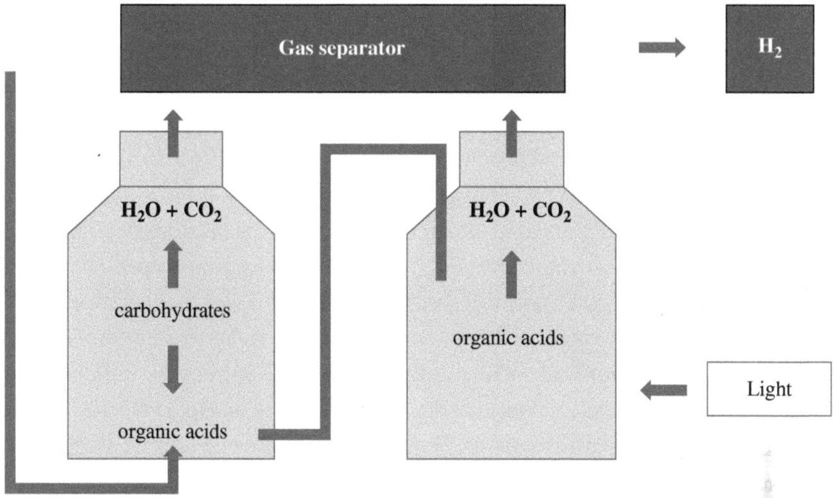

Fig. 7.12 Nonthermal H_2 production from biomass.

pellets, and briquettes) and of modern, automatically operating combustion equipment has created renewed interest in the use of solid biofuels for domestic heating. The use of liquid biofuels is also making headway for small-scale heating. For the commercial and industrial sectors, fixed-bed, fluidized-bed, and dust combustion equipment now available allow efficient production of heat from biofuels of different types and forms on a larger scale. Cofiring of biofuels with coal is also practicable. All these processes are discussed in more detail below.

Pretreatment

For domestic, commercial, and industrial combustion equipment, it is advisable to use solid biofuels that have undergone some form of pretreatment and processing (e.g., washing, drying, size reduction, and compacting) to achieve greater uniformity and ease of handling and reduce the moisture content to an acceptable level.

Small-scale heating

At the domestic level, appliances that burn wood and similar biofuels are popular because they not only provide heat, but also help create a pleasant atmosphere and decorative space. Commonly available appliances include fireplaces, heat storage stoves, pellet stoves, and burners, and central heating furnaces and boilers for wood logs and wood chips.

There is also a range of automatically operating appliances for wood chips and pellets on the market. The excellent handling properties of pellets make them a good fuel for domestic and other small-scale use. Good automatic pellet-fueled boilers with low emissions and high efficiencies have been on the market for about 10 years now. Existing oil-fired boilers can be adapted for pellet use by replacing the existing burner by one constructed to take pellets. If this burner-boiler combination is well designed, an efficiency of over 90% can be achieved, which is comparable to the efficiency of a modern oil burner. Changing an oil burner to a pellet burner—or replacing electric heating with a pellet-burning stove—can be profitable. A pellet boiler takes a little more effort to run than an oil boiler because the chimney has to be swept and the ash removed a few times a year. Apart from this, the system is completely automatic and therefore requires less effort than a traditional wood-fired boiler (Das, 2020). Environmentally, a good pellet burner is preferable to a fossil fuel burner. Pellet burners meet the demands of national and international eco-label schemes. Their emissions are well below the building regulations requirements of most EU member states and will not contribute to the greenhouse effect. A further advantage of moving from fossil fuels to biofuels for small-scale heating is that it will help the rural economy.

The use of liquid biofuels and blends of these with liquid fuels for heating domestic properties is also gaining ground. Standards for biodiesel for heating applications have just been put into force.

Large-scale combustion

The biofuels-based systems available for industrial and commercial heating can be categorized under the headings of fixed-bed, fluidized-bed, and dust combustion and cofiring (Lu, 2018).

Fixed-bed combustion includes grate furnaces and underfeed stokers. Grate furnaces, which generally have capacities up to 20 MW_{th}, are suitable for burning biomass with a high moisture content. Primary air passes through a fixed bed where drying, gasification, and charcoal combustion take place in consecutive stages. The combustion gases are burned in a separate combustion zone using secondary air (Mami, Lajili, Khiari, & Jeguirim, 2020).

Underfeed stokers, which represent a cheap and safe technology for systems up to 6 MW_{th}, are suitable for biofuels with low ash content and small particle size such as wood chips, pellets, and sawdust. The fuel is fed into a combustion chamber from below by screw conveyors and transported upward to a grate.

Fluidized-bed combustion systems are suitable for large-scale applications exceeding 30 MW_{th} in size. The biomass is burned in a self-mixing suspension of gas and solid bed material (usually silica sand and dolomite) in which air for combustion enters from below. The high heat transfer and mixing encourage complete combustion. Fluidized-bed systems allow a good deal of flexibility regarding fuels, although attention has to be paid to particle size (He, Zhang, Chen, Han, & Liu, 2019).

Dust combustion systems are suitable for biofuels, such as wood dust, in the form of small, dry particles. A mixture of fuel and air is injected into the combustion chamber. Combustion takes place while the fuel is in suspension. Fuel gasification and charcoal combustion take place simultaneously because of the small particle size. Quick load changes and efficient load control can be achieved.

Cofiring of biomass with coal in traditional coal-fired boilers is becoming increasingly popular, as it capitalizes on the large investment and infrastructure associated with existing fossil-fuel-based power systems and at the same time reduces the emission of traditional pollutants (sulfur dioxide, nitrous oxide, etc.) and GHGs (carbon dioxide, methane, etc.). Up to 10% biomass can be added to nearly all coal-fired plants without major modifications. Wood chips, willow chips, sawdust, and organic waste are the forms of biomass most often used.

Cofiring is normally realized by what is termed direct cofiring, i.e., firing the biomass and coal together in one combustion chamber of the power plant boiler. However, a number of other systems exist that fall into the category of indirect cofiring. One indirect system, known as a hybrid system, uses 100% biomass firing to generate steam. This steam is then mixed with the steam coming from the conventional coal-fired boiler for sending to the steam turbines for electricity generation. A second type of indirect system involves burning the biomass in a prefurnace and feeding the resulting flue gases into the existing coal boiler. In a third type, the biomass is gasified and the resulting combustible gas fed to the coal combustion chamber.

Emissions treatment

Appropriate techniques exist for treating all the emissions that emerge from biomass combustion plants. For large installations, flue gas cleaning is economically viable. As explained in the section on solid biofuels, NO_x emissions can usually be controlled by appropriate combustion techniques. Secondary reduction measures to remove sulfur oxides are not usually necessary because biomass combustion does not yield as much of these pollutants as does coal combustion. Solid ash and soot particles are, however, emitted by biomass combustion. As these cause aerosol formation, additional gas cleaning is required to remove them.

Electricity

The usual route for producing electricity from biomass has two stages. The biomass is converted to heat, which is then used in generation of electricity (or combined heat and power) using technology originally developed for conventional power production. A more modern approach, still at the research stage, involves the use of fuel cells. The biomass is converted to hydrogen, biogas, or methanol that is then fed to suitable fuel cells, the outputs being heat and electricity. Further information on both is given below.

Biomass-based processes for conventional power production

In the simplest route, the biomass is converted to heat by combustion, the heat used to produce steam, and the latter used to drive steam piston engines or turbines. Basic information on the conversion of biomass to heat was given earlier in this chapter. The subsequent steps involve well-proven technologies. Suitable steam engines are available for electricity production in the range from 50 kWe to 1 MWe and steam turbines for the range of 0.5–500 MWe.

The conventional steam turbine route has a number of disadvantages. For instance, the steam boiler has to be operated at high temperatures, and there can be erosion of the turbine blades due to the presence of moisture. Turbogenerators that work on the organic Rankine cycle (ORC) present a useful alternative. The ORC is similar to the cycle of a conventional steam turbine except that the fluid that drives the turbine is an organic fluid that can operate efficiently at lower temperatures to produce electricity in the range 0.5–2 MWe. The organic fluid operates in a closed cycle. It is vaporized in an evaporator by an external heat source (such as biomass-derived heat). The vapor expands in the turbine and is then condensed and pumped back to the evaporator (Campton & Rezaie, 2018).

In a development that is applicable to small-scale electricity production, heat from biomass combustion is used as the external heat source for operation of a Stirling engine. A 30 kWe prototype plant has been built that has achieved around 20% electricity efficiency in combined heat and power production.

An example of biomass-based generation of electricity by a process similar to that used in conventional power production is the straw-burning power plant that went into operation in Sangüesa, Navarra, Spain, in 2002. With an installed capacity of 25 MWe, it will, when operated for 8000 h a year, consume 160,000 tons grain straw and produce 720 TJ electricity. It was designed to burn 100% grain straw or 50% straw and 50% wood waste.

Typical production costs for biomass-based power were found to vary in 2002 between 0.088 $ and 0.252 $, depending on feedstock and country. Feedstock collection costs increase roughly with the square of the distance from the plant. For this reason, the upper size limit of a biomass-based power plant lies somewhere between 30 and 100 MWe. In general, they are below 30 MWe. This comparatively small size favors their operation as combined heat and power plants. These can meet the district heating and electricity needs of small communities (Stolarski et al., 2013).

Combined heat and power production

Combined heat and power (CHP) plants use the waste heat from electricity production for heating purposes, normally for district or industrial heating. When biomass is employed as fuel for CHP plants, the availability of a stable and sufficient feedstock supply within a reasonable distance from the plant is essential. The use of biomass in CHP plants is one of the best options for achieving simultaneously increased bioenergy utilization and significant reductions in emissions. Renewable energy sources, mainly biomass, already account for 13% of all fuel inputs to CHP in the older member states. In the new member states, the figure is just 1%, indicating that these countries have large unexplored opportunities to increase their use of biomass fuels for CHP (Colmenar-Santos, Rosales-Asensio, Borge-Diez, & Blanes-Peiró, 2016).

Fuel cell processes

A research topic that is currently receiving a good deal of interest is simultaneous production of electricity and heat by fuel cells driven by biomass-derived hydrogen, biogas, or methanol. Information on methods of converting biomass to these chemicals was given earlier in this chapter. Fuel cells require very clean fuels and are sensitive to certain substances present in biomass, e.g., sulfur. Therefore, the major problem to be resolved by the research relates to gas and methanol purification.

Liquid biofuels

The major market for liquid biofuels is in the transport sector, although products have also been developed for direct use in boilers and engines for heat and electricity production. In modern society, transport of people and goods plays a significant role in economic development and requires increasing amounts of energy.

Currently, some 98% of transport fuels are petroleum-derived. Liquid biofuels can be used neat or blended with petroleum-derived products. The possibility of using tax incentives to encourage more widespread use of biofuels has been introduced into an EU directive. The United States and Brazil remain leaders in bio-ethanol production with about 45 Mt and 24 Mt, respectively. The EU is the global leader in biomass for energy utilization and also the main importer of most biomass products (Proskurina, Jinginger, Heinimö, Tekinel, & Vakkilainen, 2019).

As a result of technological developments carried out over the past few years, transport biofuels come in a variety of types, notably bioethanol, biodiesel, and synthetic fuels (biomass-to-liquid or BTL fuel). Biodiesel is already well known. BTL, on the other hand, is at the beginning of its development.

Manufacture is from agricultural resources of different kinds—currently, grain, sugar, oil crops, etc. For the future, processes are being developed to allow lignocellulosic biomass to be a major additional source. Yields vary according to feedstock: about 1.2 tons of bioethanol can be produced from an acre of wheat and 4.1 tons from an acre of sweet sorghum. Generally speaking, production costs today are high compared with petroleum-derived products. It costs around twice as much to make a liter of biofuels compared with a liter of petroleum-derived diesel (this ratio obviously depends on the price of crude oil), and it requires on average 1.1 L of biofuel to replace 1 L of diesel. The employment prospects associated with liquid biofuels manufacturing are good: around 16 jobs per 1000 tons of biofuel can be created, mostly in rural areas (Hunsberger, German, & Goetz, 2017). The following sections discuss some of the different types of liquid biofuel in more detail.

Bioethanol

Bioethanol, a colorless liquid, is the most widely produced biofuel in the world, with Brazil and the United States being the leading producers. In 2016, world production was 18.3 million tons. In the same year in the EU, 310,000 tons were produced—some 17.8% of total EU liquid biofuels production. (The major biofuel product in the EU is biodiesel.) Despite the EU's modest production compared with other parts of the world, steady growth has been achieved in the last decade (Zabed, Sahu, Suely, Boyce, & Faruq, 2017).

Bioethanol is mostly obtained by fermentation of sugar beet, sugarcane, corn, barley, wheat, woody biomass, or black liquor. Production is generally in large-scale facilities. Most bioethanol production today is based on feedstocks from food crops. For the future, lignocellulosic biomass is expected to be an important feedstock, and its use would reduce competition between the food and energy industries for raw materials.

Because the characteristics of lignocellulosic biomass differ from those of other forms of biomass, technologies for biofuels production have to be adapted for its use. Over the past 30 years, considerable research effort has been put into this area. The focus has been on producing fermentable sugars from the lignocellulosic material that can subsequently be converted into ethanol (Singh, Shukla, Tiwari, & Srivastava, 2014).

Normal vehicles can run on a 15% blend of bioethanol and gasoline. To use pure bioethanol, they need modification. Flexible fuel vehicles adapt automatically to run on fuels ranging from pure petrol to a blend of 85% bioethanol-15% petrol known as E85. The additional costs of manufacturing such vehicles on a mass scale, compared with normal vehicle manufacture, amount to 150€ a car.

Using ethanol in vehicles is beneficial to the environment because the emissions from ethanol are cleaner than those from petrol. An 85% ethanol-15% petrol blend can reduce GHG emissions by 60%–80% compared with pure petrol. A 10% ethanol-90% petrol blend reduces emissions by up to 8%. The exact figure depends on the feedstock used to make the ethanol; a 10% blend using ethanol made from sugar, for instance, reduces harmful gases emissions by only 4%. Ethyl t-butyl ester (ETBE) is a biofuel, made from bioethanol, with an octane rating of 112 that can be blended with petrol in proportions up to approximately 17%. Methyl t-butyl ester (MTBE) has similar properties (Singh, Lakhi, et al., 2019; Singh, Tingas, et al., 2019).

Biomethanol

Biomethanol is similar to bioethanol but is much more toxic and aggressive to the engine material. It can be produced from synthesis gas made by gasification of biomass.

Biodiesel

Chemically, biodiesel consists of methyl (or ethyl) esters of fatty acids. In response to its established market for diesel engines, the EU is the principal region of the world with a developed market for biodiesel. Growth rates have been 34% a year for over a decade. Biodiesel has the largest share of the EU's liquid biofuels market: it accounted for some 79.5% of EU total liquid biofuels production in 2018. Eight member states have production facilities. Biodiesel is produced by a chemical process—the esterification of fatty acids produced from vegetable oils. Rapeseed oil is the most commonly used feedstock because as much as 1–1.5 tons rapeseed oil can be produced per acre of rape. However, sunflower oil or used cooking oils are also used as feedstock. Like bioethanol, biodiesel is manufactured in large facilities (Ballesteros & Manzanares, 2019).

Practically all diesel engines can run on biodiesel or blends of biodiesel with normal diesel. Using recently developed additives, it is also possible to blend diesel with ethanol for use in trucks. Emissions of carbon dioxide, the major GHG, are 2.5 kg/L less for biodiesel than they are for fossil fuel diesel. Emissions of hydrocarbons and soot are also lower for biodiesel than for fossil-fuel-derived diesel. In addition, biodiesel releases fewer solid particles and, because it contains no sulfur, does not create SO_2, which contributes to acid rain. NO_x emissions, on the other hand, are somewhat higher because of the presence of nitrogen in the biomass raw material.

Synthetic fuel (BTL-fuel)

BTL fuel is a short term for biomass-to-liquid fuel. Typical examples are BTL diesel and dimethyl ether (DME). BTL diesel has exceptionally good fuel characteristics—a high cetane index and low sulfur and aromatics contents. It meets all the standards for normal diesel fuel. DME is a fuel of diesel quality with physical characteristics similar to liquified petroleum gas (LPG).

An advantage of BTL fuel is its flexibility regarding feedstocks. It is produced in a two-step process involving, first, preparation of synthesis gas (a mixture of carbon monoxide and hydrogen) from a biomass feedstock and, second, conversion of the synthesis gas into liquid fuel. The biomass feedstock is converted to bio-oil by pyrolysis as described in the section on thermochemical processes earlier in this chapter. This can be carried out in decentralized units close to the place of feedstock production. The bio-oil is then gasified under high pressure (30 bar) and temperatures

(1200–1500°C) to a high-quality clean synthesis gas. Conversion of the latter to liquid fuel is carried out using a catalytic process known as the Fischer-Tropsch process originally developed to produce liquid fuels from synthesis gas derived from coal (Rimkus, Žaglinskis, Rapalis, & Skačkauskas, 2015).

Bio-oil

As indicated in the discussion on BTL fuel above, bio-oil's main use is as a valuable intermediate for production of other products. However, it also has a direct application in boilers and furnaces for heat production and in static engines for heat and electricity generation. In the future, it may have an application as a source of hydrogen.

The yield of bio-oil from the solid biomass feedstock is about 75%. Bio-oil is much cleaner than the fossil fuel-based original because it contains 100 times less ash. As a liquid, it makes a versatile energy carrier since it can be pumped, stored, transported, and burned without difficulty. Its energy density is about $20\,GJ/m^3$ compared with $4\,GJ/m^3$ for solid biomass (Hu & Gholizadeh, 2020).

Other products

Any consideration of biomass as an energy resource would be incomplete without a reference to its use as a feedstock for nonenergy products. This was touched on in the discussion of the biorefineries earlier in the chapter. A few more details are given here. A wide range of chemicals and materials can be derived from biomass. This includes the traditional plant-based products—for example, oils, starch, fibers, and drugs—for which there are already major established industrial bases. It also includes many other possibilities. For instance, lubricants made from biomass offer significant environmental advantages over their fossil-fuel-based counterparts. Printing inks, polymer additives, and polymers can also be made from biomass. Linoleum, for example, can be made from linseed oil. Surfactants are another group of products capable of being made from biomass. Some organic solvents can also be vegetable derived as can some pharmaceuticals, colorants, dyes, and perfumes. As indicated earlier, biomass has to be used wisely as an energy resource in a way that allows optimal manufacture of other priority products—not only the chemicals and materials described above, but also of food, wood products, paper and board, etc.

Wind energy

History of wind energy

Humankind has used wind power since ancient times. The application of wind energy dates back to the time of the first civilizations, when wind energy was used to power boats on the river Nile (5000 BC). In Afghanistan, the first windmills that appeared were used for grinding grain. During the 19th century, thousands of windmills were set in North America and were mainly used for pumping water on farms and plantations. At the beginning of the 20th century, small windmills began to be massively used in the United States to generate electricity, but many of them stopped working before World War II due to the intensive expansion of the electrodistributive system even to the most remote parts of the earth.

During the oil crisis in the 1970s, the production of electricity from wind recorded a sharp increase, and it was only in the 1980s that it experienced a boom upon the application of new technological achievements.

Since the 1980s, wind energy has strongly progressed in both installed capacity and in volume production. Until a few years ago, wind turbines of 500 kW were rare, but today's wind turbines with a power of 1.9–2.5 MW and a rotor diameter of 50–90 m are standard.

Wind as a potential source of energy

Wind is an inexhaustible source of energy, the global potential of which exceeds several times the world's electricity needs. The use of wind power in the production of electricity began to be used in the early 1930s of the last century when the first wind farms (plant electromechanical conversion of wind energy) began to be built. Nowadays, the wind energy industry is considered the fastest growing, both in terms of technology and in terms of growth of wind farms in the world.

Wind farms use the kinetic power of wind. Wind turbines transform mechanical energy of wind through electric generators into electricity.

About 75% of all wind farms have been installed in the EU. Out of the installed 34,466 MW, the EU serves about 3% of the total electricity needs. Germany is a leading country by the amount of installed wind power of 16,629 MW, twice as high as the strength of all power plants in Serbia, but the largest share of electricity production from wind farms is in Denmark, which creates 20% of its needs (Ortega, del Rio, Ruiz, Nijs, & Politis, 2020).

Wind power

Eliza rotor wind turbines rotate due to the air mass flow. The amount of energy transferred by the wind to the turbines is not directly dependent on the density of air, the rotor size, or the wind speed. The kinetic energy of the body in motion is proportional to its mass, so that kinetic energy depends on the density of air. More specifically, the thicker the air, the more energy is obtained on the wind turbines.

The density of air is equal to the number of molecules per unit volume of air. In normal air pressure and air temperature of 15°C, the mass of air is $1.2\,kg/m^3$, but with the increase of the moisture content, its density decreases. Colder air is denser than warm, which is why, at the same speed of wind, a wind turbine produces more electricity in winter than in summer. Atmospheric pressure decreases with increasing altitude, because at high altitudes (mountains), the pressure is lower and the air less dense.

Wind turbines convert the kinetic energy of wind in the power of the turning propeller. The surface of the rotor determines the amount of energy that can be derived from wind turbines. Since the rotor area increases with the square of the diameter of the rotor, a turbine twice as large produces four times more energy. Increasing the rotor surface is not as simple as creating long propellers, since the increase in the surface of the rotor increases the strain of the whole system, regardless of wind speed. In order to compensate for stress, the entire mechanical system must be strengthened.

Another factor that affects the input power is that wind turbines affect the turn of wind, even before the wind is affected by the rotor blades. The rotor slows down the wind speed, i.e., the wind speed is higher in front of the rotor than behind, which means that wind turbines cannot use the whole power of wind. Conversion of the kinetic wind energy into the mechanical energy of the rotating movement is done using wind turbines, which can have different structures. Modern wind turbines have wind generators with a horizontal shaft that has a system to rotate the shaft in a horizontal plane to monitor wind direction changes. Wind turbines may have a different number of blades, but, as for high powers, turbines with three blades are most commonly used. The diameter of wind turbines depends on their strength and ranges from 1 m with an output of 0.5 kW to 120 m with an output of 5 MW. The turbine drives the wind generator, which may be of a different structure. The wind turbines and wind generator, together with the pillar they are placed on, make the wind generator.

The range of wind speed during which the wind generator generates electrical energy is typically 3–25 m/s, and the maximum (nominal) power is achieved at the wind speed from 12 to 15 m/s. For wind speeds above 25 m/s, wind turbines stop due to mechanical reasons. It is not economically profitable to design wind turbines for active operation at wind speeds greater than 25 m/s, due to the fact that such winds are relatively rare. The construction of a modern wind generator is designed to withstand winds up to 280 km/h (Khosravi, Machado, & Nunes, 2018).

Wind generators are built on windy sites where the mean annual wind speed is greater than 6 m/s (at a height of 50 m above ground). One wind generator with a capacity of 1 MW can generate approximately 2000 MWh of electricity annually, enough to meet the needs of 500 average four-member households. Most of the time, wind generators are grouped together on suitable sites called wind farms. Wind farms can have hundreds of wind generators and power over 300 MW. They are constructed on land, but also in shallow coastal areas where strong and steady winds blow (see Fig. 7.13).

The price of one wind generator with a 1 MW capacity is about 1 million euros, which is equal with the cost of the installation of 1 MW in an average

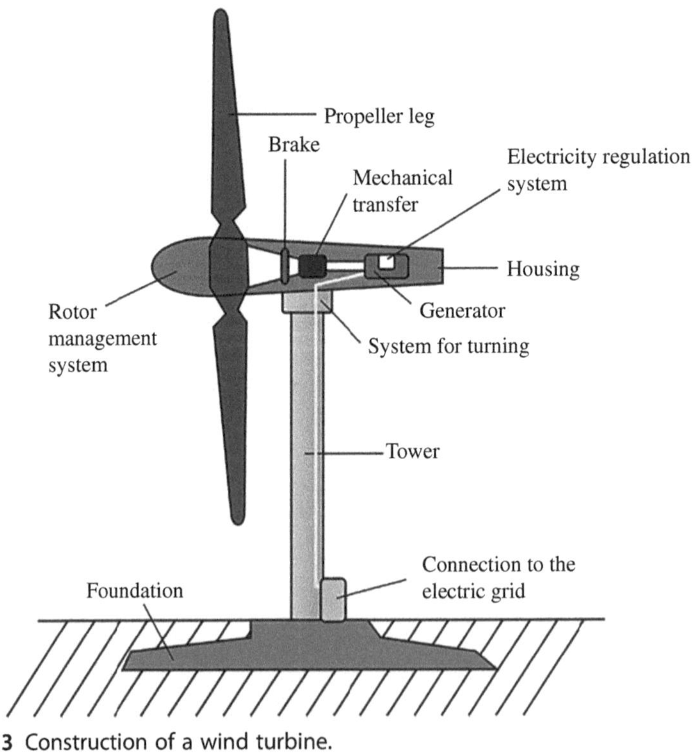

Fig. 7.13 Construction of a wind turbine.

power plant using coal. Unlike thermal and hydropower, wind farms are built very quickly, in just a few months. The life of a wind farm is about 25 years. However, the price of electricity generated from the average wind generator is still higher than the price of kWh from conventional power plants. The average wind generator with 1 MW of installed capacity in electricity generates annually twice less energy than 1 MW of installed capacity in hydroelectric power or about three times less energy than the average thermal power and about 3.5 times less energy than the same installed capacity in nuclear power. The production of a wind turbine is dictated by wind, whereas production management in a storage power plant is based on market demand. For example, reservoirs are filled at night and production is kept to a minimum, and during the day, when 1 kWh is the most expensive, the power plant operates at full capacity and produces the most expensive "current." On very windy sites, with an average annual rate greater than 7 m/s, wind farms can be more economical, but those locations are relatively rare.

It is considered that, according to the cost of produced kWh, wind farms would be fully competitive with conventional sources when the impact on the environment is included in the cost of electricity production. To produce 1 kWh in an average power plant using lignite, about 1.5–2 kg of coal is used, with the release of about 1 kg of carbon dioxide into the atmosphere and release of about 2 kWh of thermal energy that is dissipated into the environment and locally warms the river and atmosphere. Besides the water vapor, CO_2 is a leading cause of global warming (greenhouse effect). According to the Kyoto agreement, every kilogram of CO_2 that is emitted into the atmosphere has its price, and these so-called external costs can increase the cost of producing 1 kWh of electricity in power stations even up to 200% (Jansen et al., 2020).

Technology development of wind farms is intense, and it includes improvements in all aspects of their use. First, it is aimed at further increasing the power granted by the individual wind generators. Special attention is paid to the further development of wind generators for work in mountainous locations with harsh climatic conditions and turbulent winds. Some manufacturers have already successfully installed commercial turbines for extreme mountain conditions.

Construction of wind turbines

The conversion of kinetic energy into electricity is achieved using wind turbines of different structures, some of which are shown in Fig. 7.14.

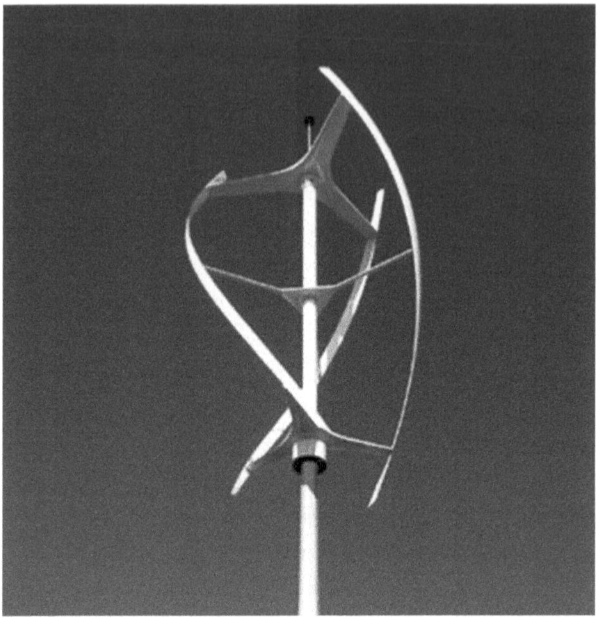

Fig. 7.14 Silent wind turbine.

According to the mutual position of rotary axes of the rotor and wind direction that runs it, wind turbines are divided into axial (axis is generally parallel to the rotor surface, i.e., the wind direction is along the axis) and radial (the rotor axis is generally perpendicular to the surface, i.e., the wind direction is perpendicular to the axis).

Axial wind turbines

Today, axial wind turbines are the most common and usually have two or three wings (Fig. 7.15). The first three-bladed axial wind turbines, designed for launching electrogenerators, were constructed in the middle of the last century in Denmark, and based on this design, most modern wind turbines of this application have been developed. Over time, airfoil wing profiles, rotor material technology, and transfer mechanisms, and today, an electrogenerator with electronic controllers, have been specialized. Electric power industries of developed countries have already disposed of axial turbines in the form of wind plantations. A single turbine already achieves the price of KWh generated electricity from 0.06€.

The length of the rotor wing, which is proportional to the torque at the output shaft, in particular contributes to more effective axial length of the wind turbines, and the risk of breaking them at wind speeds above

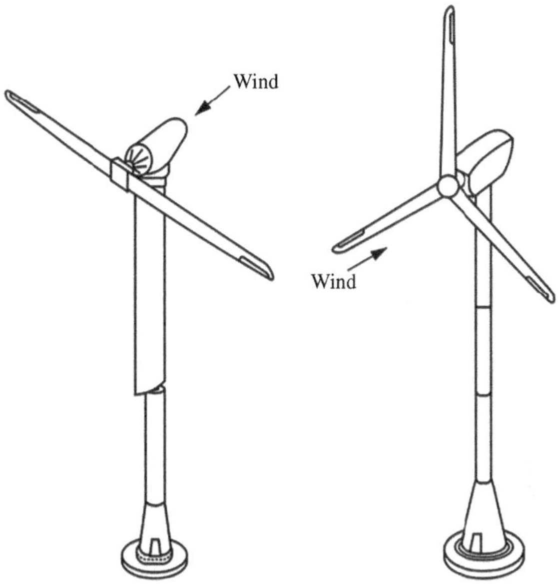

Fig. 7.15 Axial wind turbines.

15 m/s. The problem is to some extent overcome by the development of the wings from strong and light but expensive material (e.g., carbon fiber), or cheaper, but heavier materials. While the first approach can make the cost of creating them unacceptable, the second reduces work efficiency since the increase in the rotor weight slows down the adjustment of the rotor wings' position toward the change of the wind direction, and those that change direction and blow short bursts are the most unfavorable ones.

The price of active adjustment and protection (rotation of wing around its own axis and/or breaking the rotor in a horizontal plane) is only justified in these high-power wind turbines. Some specific shortcomings of axial turbines include:

- Suspension of operation in case of winds mainly above 15 m/s;
- A large area that covers rotor;
- High tower for its assembly (proportional to the length of the wing); and
- The possibility of causing dangerous (sometimes deadly) bodily harm to living beings in case of contact with the rotor in full swing.

However, the high coefficient of utilization of wind of modern axial wind turbines (50% peak and average 45%) gives them an advantage in the use for the projected strength in the range of several tens of kilowatts to several megawatts.

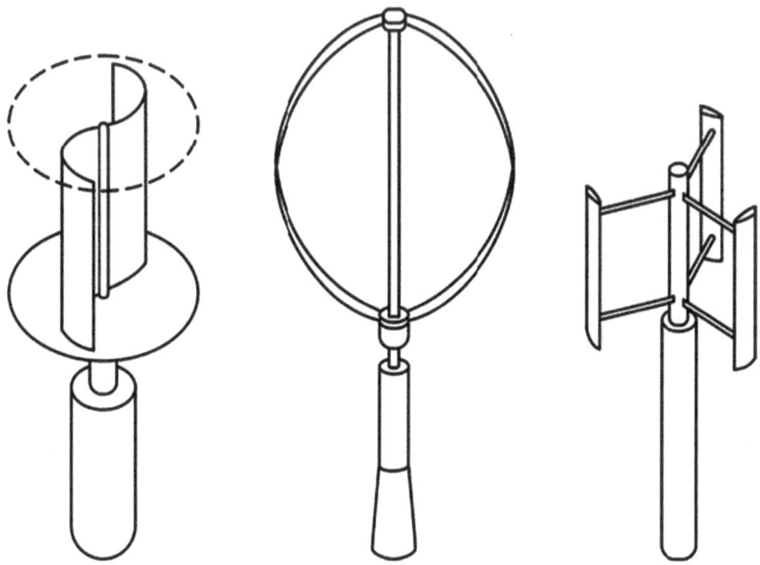

Fig. 7.16 Radial wind turbines.

Radial wind turbines

In contrast to axial wind turbines, radial wind turbines do not have to adapt the position of the rotor to the change of the wind direction, but their coefficient of efficiency is lower. The best-known types of these turbines were designed by author Sigrid Savonius (Finland), Georges Darrieus (France) (H-rotor (Fig. 7.16)), (French patent nr. 2659391).

The highest coefficient of the utilization of wind power has a Darrieus turbine (40% peak and 35% average), which may have two or more thin, slightly twisted, ribbon "C"-shaped wings. The disadvantages are that because of its balanced moment, it cannot independently start, and it occupies a large space when it is of bigger dimensions, since in that case, the shaft is strengthened by tightening ropes. This type is only found in commercial applications (the last known is 4200 kW with a two-wing rotor diameter of 100 m from Quebec, Canada, and is no longer functioning). The H-rotor turbine is similar to the Darrieus turbine, while the Riviere turbine, with its plate-like wings that revolve around its own axis. (Fig. 7.17).

The Savonius turbine, whose rotor consists of two cylindrical surfaces, partially folded and faced by a concave part, has the lowest coefficient of

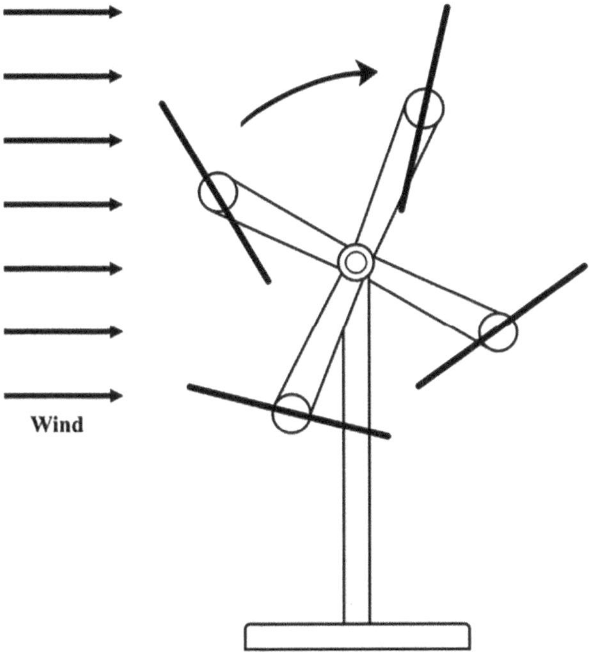

Fig. 7.17 Riviere turbine.

utilization of wind power (peak 15%), but has the advantage that it does not require a large area and cannot harm living beings that come into direct contact with the rotor working in full swing. While the axial wind turbine is close to reaching the maximum of its technical features, what has been observed in recent years is the increasing interest of specialists for underexplored possibilities of their radial cousins, and therefore new variants are frequently reported and patented.

Wind farms

Usually, on suitable sites, many wind generators are grouped and create wind farms, which may have several hundred wind turbines and power of over 300 MW. The advantage of connecting wind turbines in one location lies in the lower cost of location development, simpler interconnection with the power system, and a centralized approach to its management and maintenance. The problem that arises in connection with grouping wind generators refers to the optimal deployment of wind generators with the aim of maximizing their effects. The optimal distances are presented in Fig. 7.18, where *D* is the diameter of the wind turbine.

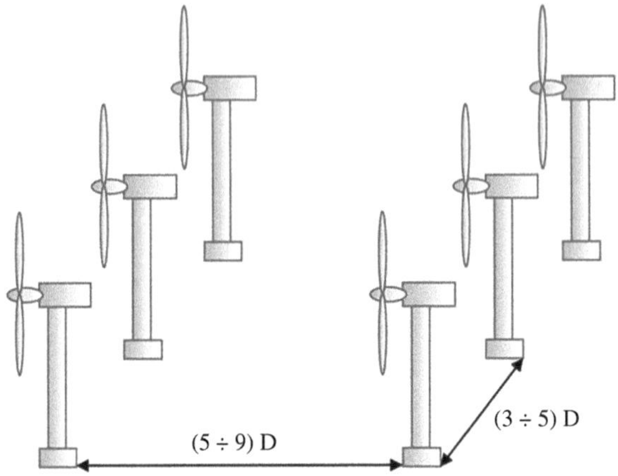

Fig. 7.18 Optimal distance between wind generators.

The annual utilization factor of a wind farm is 10%–15%. About 25% of the time during a year, plants cannot work because the wind speed is lower than the minimum and about 5% of the time, they cannot work due to the fact that it is greater than the maximum. In terms of energy, wind generators are used to fill in the base of the load diagram, and due to low power, they are usually included only in the energy balance of the local (regional) consumption (Shi, Fan, & He, 2017).

Economics of wind farms

As the starting speed threshold of the useful work of wind generators is relatively high, it is obvious that their location, given the intensity of the wind and the likelihood of windy days, is a basic factor when it comes to cost-effectiveness of their use.

Windy areas are often very remote from populated areas and from adequately developed distribution networks, and thus, the costs of their integration into the energy system influence the economy of wind powers.

The running costs of wind generators are negligible; they participate in the final price of energy with only 10% (no fuel costs, there are only maintenance costs, personnel, taxes, insurance, taxes, administrative costs, etc.). Depending on the number of windy days and the wind speed above the wind generator's useful work threshold (between 4 and 5 m/s or between 14.4 and 18 km/h), their production price (of the order 6 c/kWh during the period of 15–20 years of exploitation) is in some countries competitive to the price of energy from fossil fuels, whereas specific investments, except for gas turbine power plants, are slightly lower (about 1000–1500 $/kW) (Villena-Ruiz, Ramirez, Honrubia-Escribano, & Gómez-Lázaro, 2018).

Advantages of wind farms

Power generation using wind power has numerous advantages. First, the technology for generating power using wind energy has been known since ancient times and considerably improved since. Modern technology offers solutions that can be installed in virtually all places—in deserts, at sea, on mountaintops, in populated areas, and so on. Power generation from wind does not create any environmental pollution. This energy is produced over a system of transmission lines and thus can be transferred to places where energy is needed.

Disadvantages of wind farms

One of the most unfavorable aspects of wind farms is that their production is variable and unpredictable. This is exactly why a share of wind farms in total production should not be greater than 10%. The share of these plants could increase if accumulation of energy is ensured. Some of the possible solutions imply combining wind farms with pump-storage systems or solar power plants. Excess of electricity can be used to compress air that is then stored in underground or aboveground tanks. At a convenient time, that air can be used to run turbines, and in this way would enable nondependence in terms of time between the electricity generation in power systems and consumer spending. But none of these solutions is beneficial, because the cost of construction of other types of plants, other than the wind power plant, or costs of energy storage, can be higher than the construction of the wind farm itself.

The problems that arise in connection with the operation of wind power farms are due to moving of blades through the air, interference of electromagnetic waves, mechanical vibrations, and taking land for the construction of wind farms. Although wind turbines rotate very slowly, there are flickers of solar light, i.e., appearance and disappearance of shadows that can be very disturbing for people and wildlife living near them (Afanasyeva, Saari, Kalkofen, Partanen, & Pyrhönen, 2016).

Another particular problem is the price of energy generated by wind turbines. Technological solutions are relatively expensive and require multiyear exploitation in order to make the initial investment profitable.

Geothermal energy

Potential of geothermal energy

The word geothermal comes from the Greek words *geo* (earth) and *therme* (heat) and means heat of the earth. The heat in the earth's interior is the result of the formation of planets from dust and gases more than 4 billion years ago, and the decay of radioactive elements in rocks continuously regenerates the heat, and thus geothermal energy is a renewable source of energy. The main medium that transfers heat from the interior to the surface is water or steam, and this component is renewed in the way that water from rains penetrates deeply through crevices; it is heated there and circulates back to the surface, where it appears in the form of geysers and hot springs.

The outer rigid crust of the earth is deep, from 5 to 50 miles, and is made of rocks. The mass of the inner layer is constantly coming to the surface through volcanic vents and cracks on the ocean floor. Below the crust is a layer that extends to a depth of 1800 miles and is made from compounds rich in iron and magnesium. Below all this are two layers of the core—the liquid layer and solid layer. The radius of the earth is about 4000 miles, and no one really knows what exactly is in the interior. These are all scientific assumptions about the inside of the planet and are based on experiments in conditions of high pressure and high temperature (Miglani, Orehounig, & Carmeliet, 2018).

Down through the earth's outer layer, the crust, the temperature rises approximately 17–30°C/km of depth. Below the crust, there is a layer that is composed of partially molten rock, and the temperature of this is between 650°C and 1250°C. In the earth's core, by some estimates, temperatures could be between 4000°C and 7000°C. Since heat always moves from warmer to colder parts, the heat from the earth's interior is transferred to the surface, and the heat transfer is the main driving force of tectonic plates (Tectonic plates are pieces of the earth's crust and uppermost mantle, together referred to as the lithosphere). In places where tectonic plates merge together, there could be a failure of magma in the upper layers, whereby the magma cools and creates a new layer of the earth's crust. When the magma reaches the surface, it can create volcanoes, but mostly it remains below the surface of the vast mass of the pool where it begins to cool, a process that takes 5000 to 1 million years. Areas that are located below these magma pools have a high temperature gradient, i.e., the temperature rises rapidly by the increase in depth, and these areas are very suitable for geothermal energy.

The geothermal energy potential is enormous; there is 50,000 times more energy than all the energy that can be derived from oil and gas worldwide. Geothermal resources are found in a wide range of depths, from shallow surfaces to several-kilometers-deep reservoirs of hot water and steam can be brought to the surface and used. In nature, geothermal energy is found most often in the form of volcanoes, hot springs, and geysers of water (Lund & Toth, 2020). In some countries, geothermal energy has been used for centuries for recreational and medicinal bathing. The development of science was not limited to the area of medical exploitation of geothermal energy, but geothermal energy use is directed toward the process aimed at obtaining electricity, heating households, and industrial plants. Heating of buildings and use of geothermal energy in the process of generating power are considered the main but not the only way to use this energy. Geothermal energy can also be used for other purposes such as the production of paper, pasteurizing milk, the process of drying wood and wool, and many other purposes (see Fig. 7.19).

The main disadvantage of geothermal energy is that there are not many places in the world that are very suitable for exploitation. The most suitable areas are on the edges of tectonic plates, i.e., areas of high volcanic and tectonic activity.

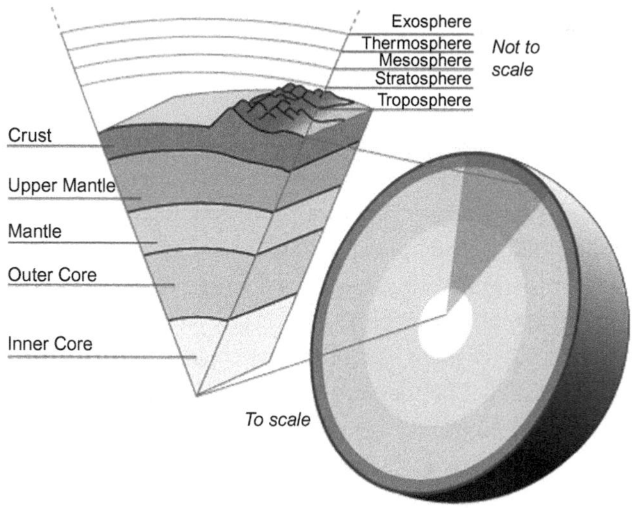

Fig. 7.19 Layers of the earth. *(From https://commons.wikimedia.org/wiki/File:Earth-crust-cutaway-english.png.)*

Electricity generation from geothermal sources

One of the most interesting forms of exploitation of geothermal energy is for the production of electricity. Hot water and steam from the earth are used to run generators, and therefore there is no burning of fossil fuels, and as a result, there are no harmful emissions into the atmosphere, only water vapor is emitted. An additional advantage is that such power can be implemented even in various environments, from farms and sensitive desert areas all the way to forest and recreational areas (Wang, Jian, et al., 2020; Wang, Zhang, et al., 2020).

The beginnings of using the heat of the earth for generating electricity are connected with a little Italian place called Landerello in the year 1904. It was in that place and that year that experimenting with this form of electricity generation started, with steam used to run a small turbine that fed five lightbulbs. This experiment is considered the first use of geothermal energy for electricity generation. There, in 1911, construction of the first geothermal power plant started, which was completed in 1913, and its power was 250 kW. This was the only geothermal power plant in the world for almost half a century. The working principle is simple: cold water is pumped to the hot granite rocks located near the surface, hot steam comes out at over 200°C under high pressure, and the vapor then drives a generator. Although the plant in Landerello was destroyed in World War II, it was rebuilt and extended and is still in use today. This plant is still powered by electricity that supplies about 1 million households, i.e., produces almost 5000 GWh/year, which is about 10% of total world production of electricity from geothermal sources (see Fig. 7.20).

Types of geothermal power principles

The principle of dry steam—only hot steam is used, typically above 235°C. This steam is used for direct setting of turbine generators into motion. This is the simplest and oldest principle and is still in use today since it is by far the cheapest process of generating electricity from geothermal sources. The Landerello geothermal power plant just discussed used this principle. Currently, the largest power plant using the "dry steam" principle is in Northern California and is called The Geysers. It has generated electricity since 1960, and the amount of electricity produced from this plant is still sufficient to supply a city of San Francisco's size (Manzella et al., 2018).

Flash principle (flash steam)—hot water from geothermal reservoirs, which is under great pressure and at temperatures above 182°C, is used. Pumping water from these reservoirs toward power on the surface reduces

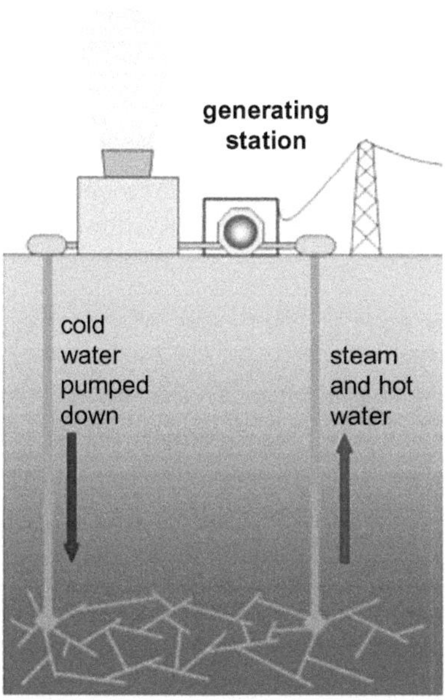

Fig. 7.20 Principle of generating electricity from geothermal sources.

the pressure, and the hot water is turned into steam that sets turbines into motion. The water that did not turn to steam comes back to the reservoir for reuse. The majority of modern geothermal power plants use this principle.

Binary principle (binary cycle)—water used in the binary principle is colder than the water used for other principles of generating electricity from geothermal sources. In this procedure, hot water is used as heating fluid that has a much lower temperature than boiling water, and the liquid evaporates at a temperature of hot water and sets turbines of generators into motion. The advantage of this principle is the higher efficiency of the procedure, and the availability of geothermal reservoirs is much higher than in other procedures. An additional advantage is the complete isolation of the system, since the used water flows back into the tank and therefore the heat loss is reduced, and there is almost no water loss.

The choice of procedures to be used in the construction of new power plants depends on the type of geothermal energy sources, i.e., temperature, depth, and quality of water and steam in the selected region. In all cases, the condensed steam and geothermal liquid residues are returned back into the bore, thus increasing the durability of geothermal resources (Rosen & Koohi-Fayegh, 2016).

Heat generation from geothermal sources

Another form of using geothermal energy is heat. The largest geothermal system used for heating is located in Iceland, in its capital city of Reykjavik, where geothermal energy is used in almost all buildings, and as much as 89% of Icelandic households are heated in this way. Although Iceland is plausibly the largest beneficiary of geothermal energy per capita, it is not alone in the field of geothermal energy use.

Geothermal energy is also used in agriculture to increase yields. Water from the geothermal reservoir is used for heating greenhouses for the production of flowers and vegetables. The heating of greenhouses does not imply only heating of air; the soil in which plants grow is heated as well. For centuries, it has been used in central Italy and Hungary, and currently, it covers 80% of energy needs for greenhouses heating.

Heat pumps are another way of using geothermal energy. Heat pumps use electricity to circulate the geothermal fluid, a fluid that is later used for heating, cooling, cooking, and hot water, thus significantly reducing the need for electricity (Noorollahi, Saeidi, Mohammadi, Amiri, & Hosseinzadeh, 2018).

Since the estimated total quantity of geothermal energy that could be used is much higher than the overall amount of energy sources based on oil, coal, and natural gas together, geothermal energy should be given more importance because it presents a cheap, renewable source of energy that is more environmentally friendly. Since geothermal energy is not readily available everywhere, at the very least it should be used in the places where the energy is readily available (edges of tectonic plates) to reduce at least a little pressure on fossil fuels and thus help the earth recover from harmful GHGs.

Hydropower energy

Ocean energy

Oceans cover more than 70% of the earth's surface and thus represent a very interesting source of energy that in the future could provide energy to households and industrial plants. Currently, ocean energy is an energy source that is very rarely used since there are a small number of power plants using ocean energy, and the plants are still small in size (Melikoglu, 2018). But as a renewable sector it is becoming more important. There are three basic types of solutions that are used in the exploitation of ocean energy:

- The energy of waves;
- Ocean tidal energy; and
- The thermal energy of the sea.

Wave energy

Wave energy is a form of kinetic energy that exists in the motion of waves in the ocean. Wave motion causes the winds blowing over the surface of the ocean. This energy can be used to run turbines, and there are plenty of places where the winds are strong enough to produce a continuous wave motion. Huge amounts of energy lie in wave energy, offering enormous energy potential. Wave energy is captured directly beneath the surface of waves or from various fluctuations of pressure below the surface. This energy could drive a turbine: the wave rises in the chamber and the growing power of water forces air from the chamber and then the movable air drives the turbine, which then powers the generator.

The main problem with the energy of waves is the fact that this energy source cannot be used uniformly in all parts of the world; many research projects are dedicated to solving this problem of uniformity. But there are also many areas with very high levels of utilization, such as the west coast of Scotland, Northern Canada, Southern Africa, Australia, and the northwestern coast of North America (Leijon & Boström, 2018). There are many technologies for the exploitation of wave energy, but only a few are actually commercially exploitable. Technologies for the use of energy waves are not only installed at the coast, but also far out on the open sea, with systems set installed in deep water, at depths exceeding 40 m (see Fig. 7.21).

Most technologies for the exploitation of wave energy are still focused on the closeness of the coast, or are on the coast, and the difference between them lies in their orientation toward the waves with which they interact, and in the working principle by which the wave energy is converted into the desired shape of energy (Zheng, Wang, & Li, 2017). So-called "terminator devices, point absorbers, attenuators, and

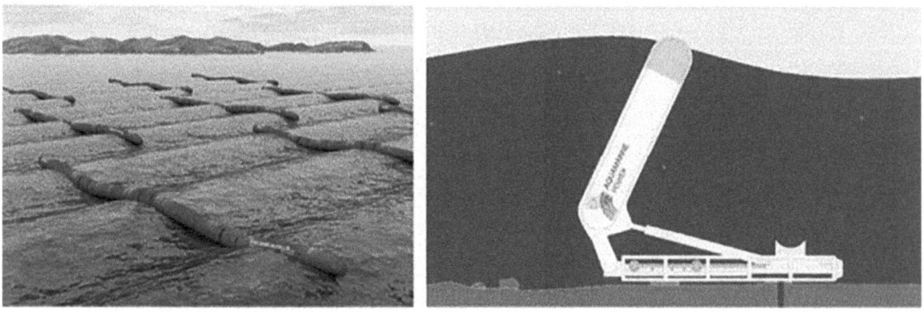

Fig. 7.21 Use of wave energy.

overtopping devices" are certainly among the most popular technologies. "Terminator devices" such as "oscillating water columns" are usually somewhere on the coast or near the coast, spreading perpendicularly with regard to the travel direction of a wave, whereby after the wave power is captured and reflected, the "oscillating water column" is then like a piston moving up and down, forcing air through a hole connected to the turbine. "Point absorbers" are different types of technologies that include a floating structure with components that move in relation to each other due to wave energy and then create energy since the movement pushes electromechanical or hydraulic energy converters. Attenuators can also be floating structures that are oriented parallel to the direction of the wave, where the difference in the heights of waves varies according to the length of the device that causes bending in places where the parts merge together, and this bending is coupled with a hydraulic pump or other converters for further transformation into useful forms of energy (Khan, Kalair, Abas, & Haider, 2017).

"Overtopping devices" are actually reservoirs that are filled with the coming waves to levels above the average of those from the surrounding ocean; after the water drains, the gravity forces them back toward the ocean surface, and then that energy drives water turbines. Although the potential of wave energy is unquestionable, there are certain aspects that should be taken into consideration, especially problems concerning the surrounding environment.

Tidal energy

The amplitude of tides depends on the mutual position of the sun, moon, and earth, and the amplitude and frequency are different on some coasts. The amplitudes of the tides in the Mediterranean Sea are 10 cm, and in the Atlantic, Pacific, and Indian oceans, they are approximately 6–8 m. In some parts of the coast in western France and in the southwestern part of Great Britain, the amplitude reaches more than 12 m. On the West European Atlantic coast, the time gap between the two tides is 12 h and 25 min, and on the shores of Indochina there is only one tide in 24 h (Devlin, Pan, & Lin, 2018).

For energy utilization of the tides, it is necessary to select a suitable place on the coast, with large amplitude of the tide, as well as with the possibility to isolate a part of the sea surface (constructing the partition) to create the storage pool. The simplest way to achieve this is by building turbines that operate in only one direction of water flow.

In order to prolong the time of operation, a turbine operating in both directions of water flow can be set up: from the pool and into the pool. There is a variant of a turbine that works like a pump, either to transfer water from the pool into the sea or from the sea into the pool. In this way, the potential energy of the tide is better exploited.

However, regardless of the selected mode of exploiting tidal energy, neither uninterrupted drive nor constant power can be achieved. This shows that the production of electricity on this basis is impossible in isolation, without cooperation with other plants for generating electric energy. Such plants must be included in the power system in which other plants have a total force of several tens of times greater than the power of these plants. For the plant to be economically viable, it must fulfill two requirements: the length of the dam should not be too long and the depth of the dam site should not be too deep.

The total tidal energy is estimated at 26,000 TWh or 2230 Mtoe in a year. A third of this is lost in the shallow seas. The mean amplitude of oceanic tide widths is less than 1 m, and the energy utilization of tides is economically justified if the amplitude is greater than 2 m. Therefore, it is estimated that out of the total energy of tides, only 2% is usable, which is 520 TWh/year, about 3% of the current annual production of electricity (about 15,000 TWh). The actual electrical energy is even smaller, because even through the best sizing only 25% of theoretical production can be used (Neill, Hashemi, & Lewis, 2016).

A tidal energy plant was built in France, called the La Rance plant, which was set into motion in 1966. It has 24 turbines that can operate as pumps for both directions of water flow. The total power is 240 MW turbines. The annual production is 608 GWh (20% of the theoretical production), and 64 GWh is required for pumping (the energy is taken from the system). The usable pool capacity is 184 million m^3, the surface area is 22 km^2, and the length is about 12 miles. The length of the wall is about 720 m. In Russia, not far from Murmansk, an experimental plant was built (in operation since 1968) with a power of 800 kW.

This type of power requires high investment costs, and the total possible production of electricity on favorable locations is only a marginal amount of the energy needed.

Thermal energy of the sea

Ocean thermal energy conversion is a method used to create electricity that uses the temperature difference that exists between deep and shallow water, since the water is colder at greater depths. The greater the temperature difference, the greater the efficiency of the whole method, and the minimum temperature difference should be 38°C. This method has a long history of use and dates from the early 19th century. Most experts believe this method is a feasible investment and even with the existing technology, a very large amount of electricity could be produced (Chen et al., 2019). But this is not the case today as OTEC (ocean thermal energy conversion) requires huge, expensive large-diameter pipes that must be set at least a kilometer into the sea so that they can bring colder water from greater depths, which is of course very expensive. The types of OTEC systems include (Wu et al., 2020):

- Closed-cycle systems that use liquids with low boiling points, usually ammonia, and thus drive a turbine, which then generates electricity. Warm surface seawater is pumped through a heat exchanger, and because of the low boiling point, it evaporates, and the generated vapor then runs the turbo generator. The colder deeper water is then pumped through a second heat exchanger where, due to condensation, it passes through the vapor back into liquid, and the liquid is then recycled through the system. In 1979, the Natural Energy Laboratory, in collaboration with several partners, created a mini OTEC experiment, which was the first successful OTEC system of a closed sea that was constructed in the sea. The mini OTEC vessel was transported 1.5 miles from the Hawaiian coast, and it managed to produce enough energy to power lights on the vessel and to ship computers and TVs. Twenty years later, in 1999, the Natural Energy Laboratory tested a pilot plant of a closed system, with power of 250 kW, the largest plant of this type ever put into operation.
- Open-cycle systems that use warm surfaces of tropical oceans for getting electricity due to the fact that hot water boils after being placed in a container with low pressure. After that, the expanding steam begins to force a low-pressure turbine connected to an electrical generator, and eventually, it condenses back into liquid due to exposure to cold temperatures in the deep ocean.
- Hybrid systems that are designed in a way that combines the positive effects of both of open and closed systems. Operation of hybrid systems includes warm seawater that enters the vacuum chamber where it is converted to steam (a process similar to the one in the case of open systems). After that, the vaporization of steam into a low boiling liquid (as in closed systems) then drives a turbine and creates electricity (see Fig. 7.22).

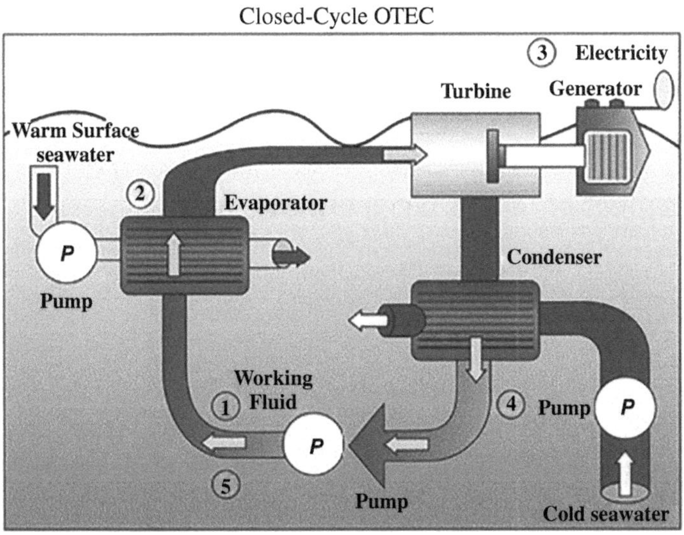

Fig. 7.22 OTEC system.

OTEC has a lot of potential to generate electricity, but electricity is not the only positive benefit of using an OTEC system. As a by-product, cooling of the air can be produced, and the used cold seawater from the OTEC plant can either chill freshwater in the heat exchanger or flow directly into a cooling system. Additionally, some types of fish, salmon, for example, can reproduce much better in deep water rich in nutrients, a product of the OTEC's work. But since OTEC plants require a large initial investment, create certain environmental issues, and require a lot of building space, these systems are not ideal. Another factor affecting the commercialization of OTEC projects is the fact that the world has only a few hundred sites suitable for construction, mostly in the tropics, where the deep ocean is rather near the coast, and thus the additional costs that would be incurred during the construction of an OTEC project away from the coast are avoided (Zhang, Xu, Song, Huang, & Wu, 2019).

Ocean energy is a renewable source of energy that definitely requires more research, mainly in order to increase investment effectiveness and reduce the huge start-up costs, which is the biggest disadvantage of this renewable source. The oceans represent two-thirds of the earth's surface and as such, have huge potential worthy of further research. However, current technologies do not support the profitability of these projects. More research needs to be done. Problems related to the massiveness of these plants or return on investment are certainly highlighted, but they are not the only issues, since it is also necessary to meet certain environmental standards before entering into larger projects, as to preserve the environment to the greatest possible extent.

Management of renewable energy sources

The impact of renewable energy on the environment is project specific for each site, but generalizations are still possible. Renewable energy sources tend to be more on the side of environmental protection in comparison to the energy that comes from conventional sources, especially in terms of emissions to air, and the lifecycle of emissions from renewable energy sources is insignificant in comparison to that of fossil fuel plants. Renewable energy projects should be considered in the framework of sustainable development that integrates energy demand reduction and energy efficiency, using a mixture of energy from various renewable sources to meet the increased scale of energy demand and, on the other hand, the protection of biodiversity and local communities (Hicks & Ison, 2018).

Technologies for renewable energy sources vary in terms of efficacy and acceptability of environmental protection. Therefore, each individual project that is planned should be evaluated in terms of finding alternatives to find the most appropriate technology for each particular case. In the development phase of the project, it is necessary to give equal importance to social and environmental, as well as technical, economic, and financial aspects, whereby the assessment solutions should include all political, program, and project options. Strategic assessment of environmental impact and life cycle analysis should be integrated and done as an initial step in the process and should give equal importance to the options both on the demand side and on the supply side. Renewable energy must be supported to enter the energy market, but this support must be limited to renewable energy produced in a sustainable manner (Maqbool, 2018).

Hydropower

Small and mini hydropower plants, which produce between 100kW and 10MW of electricity, often produce enough electricity to be integrated into the power grid. These plants do not require accumulations and do not disturb the flow of a river or stream, and they can be very effective in energy supply to networks in those areas where there are rivers or waterfalls (Couto & Olden, 2018).

Large hydroelectric power plants cause problems when it comes to regulating water courses, damaging landscape, impacting flora and fauna, the emission of GHGs (methane which is released from the flooded area), affecting the quality of drinking water (changes in dissolved substances in water and oxygen level, presence of toxic substances, changes in temperature and acidity, turbidity, etc.), and creating noise and negative visual impact for residents in the area. Large dams can cause the relocation of hundreds of thousands of people, destroying their lives and communities. Ensuring adequate compensation, moving, and reconstruction can prove be very difficult and, in many cases, communities affected by such projects remain in a worse position than they were before the project (De Faria, Davis, Severnini, & Jarmillo, 2018).

The World Commission on Dams estimates that 1%–28% of the total world emissions of GHGs come from hydroaccumulation. In some cases, emissions from accumulations may be equal to or greater than those from power plants working on coal or gas, and emissions are the highest in shallow, tropical accumulations (Zhang & Xu, 2015).

The hydrological cycle is renewable, but large hydropower plants do not use only water stream as a drive, but hydroaccumulation, which are reduced in some cases very rapidly (accumulations around the world are losing capacity upon the sediments increased at an average rate of 0.5%–1% per year).

The criteria for the realization of getting energy from water courses include:

- Project is less than 10 MW;
- Project does not include a dam, accumulation, and resettlement;
- Project in no way affects the regime of watercourses and movement of wild animals;
- Project does not affect biodiversity or human needs for water; and
- Project does not affect the possible investment in the renewal and increase in efficiency of existing energy units in the project area.

Solar energy

Solar energy, including concentrated systems using reflective materials such as mirrors to concentrate solar energy and then convert heat into electricity, is becoming increasingly cost-effective for the supply of electrical networks, although it is still far more expensive than alternative technologies. Solar PV systems, which directly convert sunlight into electricity, are particularly suitable for rural electrification in areas that do not have the conditions for microhydropower plants (Kannan & Vakeesan, 2016). Such systems can be used for power generation, water pumping and maintenance, health systems, and communications. PV systems have few negative effects during use, but the production of their cells requires careful control of the use of potentially toxic and hazardous materials (Tammaro, Salluzzo, Rimauro, Schiavo, & Manzo, 2016).

The use of solar collectors for generating heat is a far more cost-effective option. It is estimated that the costs of heating and water heating are more than 60% higher than the total cost of energy for a single building. It is possible to cover 50%–65% of annual hot water needs with solar energy and the appropriate-sized system. The overall need for hot water in summer in most cases can be met with solar heating systems. Then the conventional system can be completely excluded. This is particularly advantageous because in summer, the system operates with a low level of capacity utilization due to lower demand for heating (Jamar, Majid, Azmi, Norhafana, & Razak, 2016).

So far, significant impacts of solar technologies on the environment or society have not been identified. However, land-related issues can be raised—when selecting locations for plants, agricultural land must be avoided. The potential impact on nature should be taken into consideration.

Wind power

Wind power is generally a cheaper option than solar energy in locations where the average wind speed is greater than 4 m/s during the weakest windy weather. Although there are doubts about constancy of wind, they can be solved by combining wind with other renewable sources such as solar energy and hydropower. Wind power plants have very low emissions throughout their life cycle, but have a number of consequences for the environment that may reduce their potential. The most important consequences are (Gao et al., 2019):

- Visual effect—wind turbines are too conspicuous, since they must be installed in prominent places. This (reasonably) can be considered a violation of the landscape, and dissatisfaction on this point grows along with the increasing dimension of the new generation of turbines.
- Noise—wind turbines generate aerodynamic noise from wind passing over the blades and mechanical noise from the moving parts of turbines, particularly from the housing of the generator. However, improvements in design to some extent reduce the noise. Turbine farms built away from densely populated areas are generally less irritating.
- Electromagnetic interference—wind turbines can dissipate electromagnetic signals causing interference to communication systems. Placing in the proper location (avoidance of military training grounds or the airport) can reduce this impact.

- Security of birds—birds can be hurt by contact with rotating turbine blades. Migratory birds are more vulnerable than species that are not moving. Placing turbines outside the path of migratory birds reduces this impact.

The criteria for the realization of getting energy from wind power include (Gao, Guo, Ma, Huang, & Li, 2020):

- Project is not running in protected nature areas;
- Project is not along the migratory paths of migratory birds;
- Project has no impact on the population of bats (in addition to problems of contact and disruption of habitat, also the issue of ultrasound emissions needs to be addressed);
- Projects of wind turbines to be based on the studies of biodiversity and to be carried out along with impact assessment on the environment, like any other industrial project;
- Projects on wind energy production should be supervised to ensure they do not create a negative impact on human communities and wildlife;
- Project should use the latest equipment in order to eliminate noise, vibration, and electrical and magnetic fields, and international financial institutions should not support the usage of old, used installations; and
- Projects of coastal wind power plants should be based on accurate analysis of potential impacts on birds and mammals, including their habitat and feeding areas and resources.

Geothermal energy

Geothermal energy is a clean, renewable energy source from which the world can get heat and electricity. It is considered a renewable source after the heat is released inside the earth and is essentially unlimited (Manzela, 2018).

Geothermal energy can be used to produce electricity for immediate use and for home heating (through geothermal heat pumps). Geothermal energy relies on an existing, permanent source of heat for power generation and therefore is considered a base, constant supply. Since some renewable energy sources can be used only under favorable weather conditions, they are considered to have limited availability to meet the constantly growing needs of the 21st century. However, geothermal energy has the potential to provide reliable sources of electricity with significantly lower levels of emissions from fossil fuels, and it eliminates the problem of radioactive waste disposal. The factor availability of geothermal energy is about 95%, which means that geothermal power plants can be used during 95% of any period of time, based on many decades of observation of these plants. The capacity factor of geothermal energy ranges from 89% to 97%, depending on the type of system that is in use (Pan et al., 2019).

Geothermal energy during operation may discharge emissions into the atmosphere. As for the gases, they are primarily carbon dioxide and hydrogen sulfide, along with traces of ammonia, hydrogen, nitrogen, methane, and radon, and volatile metals such as boron, arsenic, and mercury. Emissions should be controlled by strict regulations and control methods of the geothermal industry itself used to check the compliance with the requirements of this regulation. Systems of decreasing hydrogen sulfide reduce damage to the environment but are expensive to install (Soltani et al., 2021).

The criteria for projects in this area include (Shortall, Davidsdottir, & Axelsson, 2015):

- Project relies on the return of the used geothermal water back into the earth, so there are no possibility of contamination of river or lake systems with hot water; and
- Process equipment is used for the elimination of hazardous emissions of GHGs, hydrogen sulfide, and other gases from the thermal waters.

Biomass

Biomass is a renewable source of energy that occurs in various ways, both by people and by nature. It is obtained from many sources, including from by-products of the wood industry and agricultural crops. Biomass does not release carbon dioxide into the atmosphere after it absorbs the same amount of carbon during the growth that is released during use (but the energy consumed during transportation and processing should be taken into account). The advantage of biomass is that it can use the same equipment to produce electricity in existing power plants that burn fossil fuels. Biomass is an important source of energy and the most important fuel worldwide after coal, oil, and natural gas (Thrän, Seidenberger, Zeddies, & Offermann, 2010).

Production of biomass for energy purposes implies the use of a large area, which, when connected with the usual method of land, creates a significant impact on biodiversity and the manner of its production. Therefore, the use of crop residues to produce electricity, heat, and biodiesel, whether from sugar beet, rice chaff, or straw and other waste from crops, forestry activities, or vegetable oil, etc., is one of the best ways of sustainable energy production, while it does not preclude other important ways to use agricultural waste such as soil conservation, for example. Obtaining methane for energy production from the decomposition of municipal solid waste (for sanitary landfills) or agricultural waste (e.g., pork and poultry manure) may be a viable alternative, depending on the method of treatment of these wastes. Burning—incineration—of municipal and industrial waste is not an acceptable solution, because it creates a wide variety of polluting emissions, which are very dangerous to human health and the environment.

Decisions about acceptance of burnable waste are less sustainable and represent a bigger threat to the society compared with other alternatives for solid waste management.

- Solid biomass: Organic, nonfossil material of biological origin that can be used as fuel for heat or electricity.
- Wood, wood waste, other solid waste: Dedicated crops grown for energy (poplar, willow, etc.), a number of materials from wood obtained in industrial processes (especially in the wood and paper industry) or indirectly in the forestry and agriculture (burnt wood, waste branches, wood chips, sawdust, wood chips, resin, etc.), and other wastes, such as grass, rice chaff, shells of nuts, husk, poultry waste, and others.
- Dedicated crops: In theory, any plant material can be used for bioenergy production, but the material that is grown specifically for this purpose is characterized by large amounts of biomass and high energy potential. Bioenergy can take many forms, including bioelectrical energy and biofuels. Biogas and biodiesel fuels are derived from plants, which can be used in much the same way as natural gas or petroleum. When the biofuel is produced from domestic plants or plants grown without fertilization and without irrigation, it creates significantly less emissions than conventional petroleum. If they are produced and used locally, biofuels can help the local economy.

Biofuels are a comparatively clean alternative to oil as a fuel source and can be particularly useful for traffic. They have the potential to provide a fuel that emits small amounts of carbon compared with the amount of conventional fossil fuels. Also, biofuels generally produce fewer polluting emissions than other fossil fuels. However, one must carefully consider the impact of production and the use of biofuels on the environment in a broad sense, including the impact of biofuel production on local ecosystems. For example, the planting of forests as plantations for biofuel is problematic, since such plantations require enormous amounts of water or the use of dangerous supplements in fertilizers and pesticides. The amount of conventional fuel needed for the production of certain biofuels cancels or nearly cancels the usefulness of biofuels in reducing the amount of GHGs. Biofuels must be accepted, with caution, as an alternative fuel source since they are one of the few alternatives to conventional oil use (Correa et al., 2019).

Biodiesel is a fuel whose properties are the same as those of mineral diesel, but it is derived from renewable energy sources. In Great Britain, the most attention has been focused on the production of biodiesel from rapeseed oil. The technical term for the fuel from seed is methyl ester. Any cooking oil, including used (fried) oil, generally has the potential to produce biodiesel. Crude oil from rapeseed goes through the esterification process that removes glycerin, allowing oil to have properties similar to mineral diesel. Vehicles do not require any modification to use biodiesel, since it can be mixed with normal diesel. There are no significant differences found in engine performance upon the comparisons conducted between biodiesel and conventional diesel (Roy, Wang, & Alawi, 2014).

Biodiesel is significantly safer than diesel generated from oil since it has a lower flash point, it is more difficult to ignite, it does not produce explosive gases, and it even has a lower degree of toxicity for humans and animals if swallowed. Biodiesel is biodegradable, and in case of spillage, less damage is created to the environment. Usefulness of the program includes reduced levels of carbon monoxide and total hydrocarbons and mild odor. Contrary to conventional diesel, it is completely free of aromatic compounds and sulfur, which are poisonous and subjects to legislation.

The criteria for biomass energy projects include (Famoso, Prestipino, & Galvagno, 2020):

- Planning and planting promote the protection, restoration, and conservation of natural forests and do not increase pressures on natural forests and protected nature areas;
- Plantations do not have a negative impact on natural habitats;
- Use of genetically modified organisms in crops is excluded;
- Native species are preferable to exotic species in plantations and restoration of degraded ecosystems. Exotic species, which should be used only when their properties are better than native species, should be carefully monitored to detect unusual mortality, disease, or excessive multiplication of insects and other adverse impacts;
- Project improves the composition of soil, fertility, and biological activity;
- Project does not involve the use of dangerous fertilizers and insecticides;
- Project does not have negative impacts on water availability and its quality or impact on river and lake systems in terms of water;

- A single plant species is not planted on large areas, unless tests and experience demonstrate that it is ecologically suitable for it, that it is not invasive, and it does not have significant negative ecological impacts on other ecosystems;
- Project does not raise issues of land ownership, use, or access;
- Project does not jeopardize the safety of food at any level (plantations for energy dramatically reduce/eliminate food crops in the area where they are present);
- Project does not include the increase in GHG emissions;
- Biomass as an energy source should be of domestic origin (no import of biomass from third-world countries);
- Project cannot give rise to social conflicts; and
- Production of biomass must have an obvious positive balance/energy balance (the produced energy against the all energies spent in the process).

Biogas is a gas that comes from the decomposition of biological (organic) waste regardless of whether the decomposition takes place at the landfill, in a closed equipment for the anaerobic decomposition of plants, or in wastewater treatment facilities. Utilization of this atypical source of waste gas is not only beneficial for the environment, but also for owners of milk farms. Although it requires the use of digesters, animal waste can be converted into valuable liquid natural gas, fiber, and fertilizer (Berendes, Yang, Lai, Hu, & Brown, 2018). Biogas, whether generated as a by-product of the digester processing brewery hops or more often from the decomposition of animal waste, is an excellent source of energy after the predominant methane gas is purged. Liquefied natural gas obtained from biogas can be a locally produced, clean, sustainable, and cheap source of renewable energy, which makes it a very attractive option for growers on farms for milk production and end users of gas in the local community. A direct negative impact has not been identified. The recommendations are possible in terms of decentralization—local use of biogas, for example.

Fuel cells

Fuel cells are electrochemical devices that convert fuel energy directly into electricity. They function like regular batteries when they are fed with the fuel (hydrogen) through the anode and an oxidant (e.g., air) through the cathode. Fuel cells bypass the traditional production of thermal energy in the form of combustion, the conversion of heat into mechanical energy (as with the turbine), and ultimately convert mechanical energy into electricity (e.g., using the generator). In contrast, fuel cells chemically combine the molecules of fuel and oxidant without burning, thus freeing inefficiencies and pollution of traditional combustion (Sharaf & Orhan, 2014). Hydrogen can be produced by electrolysis using energy generated from renewable sources or hydrogen-rich fuel transformation such as oil and natural gas. The process of reshaping transmits relatively small amounts of CO_2, but it does not produce emissions of other pollutants generated by burning fossil fuels. Technology is at a relatively early stage of development and not yet cost-effective compared with the price of conventional energy sources. Possibilities of future use of fuel cells are, however, very broad and include transportation (private vehicles, public transport) as well as domestic and industrial needs (Munoz-Cupa, Hu, Xu, & Bassi, 2021).

Nonfinancial effects of renewable energy

Improvements in energy management are the consequence of long unplanned exploitation of natural resources and pollution of the atmosphere as a consequence of their exploitation and use. The energy problem is globally recognized as one of the priorities of sustainable development that can and must be solved by applying different activities at all levels. For all projects related to quality management of sustainable energy development, there are large groups of interested entities including (Klein & Coffey, 2016):

- Companies—as places of production or consumption, through which the application of the appropriate management methods acquires a number of advantages that are reflected in the positive image acquisition, financial gain, lower costs, and the provision of certain subsidies;
- Users—as consumers of the goods or services of companies who wish to support the energy efficient companies and therefore produce very strong pressure on the responsible institutions;
- Citizens—as consumers of energy, who are encouraged to save energy and to independently invest in renewable energy sources;
- Legislation—as necessary framework for the functioning of the system, has to regulate the area of sustainable energy management and to be adjusted depending on changes in the demands of global integration;
- Investors—as a part that wants to direct their capital toward companies that do environmentally responsible business, investing in projects aimed at energy saving or production of energy from renewable sources;
- International community—this group is interested in efficient energy management and energy stability after one of the key issues for the survival of the world community;

- Current generations—as a party that has shown interest in the regulation of the complex issue of energy management, regardless of the fact that full effects of the implementation of projects in this field cannot be expected in the short term; and
- Future generations—as the biggest beneficiaries of changes in the safe area of production and consumption of energy, who are entitled to have the same amount of resources that the generations before them disposed of, which is a key principle of sustainable development and the principle of intergenerational justice.

Economic effects of sustainable energy management

Sustainable energy management is one of the key demands of managing natural resources and sustainable development in general. Sustainable development, in essence, represents a system of our integrated subsystems (economic, environmental, social, and institutional), and any change in one subsystem causes effects in the remaining three. Therefore, no changes or activities should be observed independently and separately from the others (Lee & Cheng, 2016).

Special problems exist in balancing economic and environmental development, as they are largely interconnected and often directly opposed. Economic development has always been the only indicator of development and is measured by specific economic indicators. The development measured in this way did not take into account the impact that economic development had on the environment, as measured primarily by resource depletion and pollution production. Economic development has had its price, which was not considered until sustainable development concept set the criteria and caused a reconsideration of the impact of economic development of the environment.

The modern concept of sustainable development questions the justification of economic development at any cost, and in the process of measuring the development level, it introduces new indicators that are directly opposed to them. Indicators of environmental quality and quality of life are becoming equally important parameters. In this way, economic development ceases to be the dominant goal (Moustakas, Loizdou, Rahan, & Nizami, 2020).

Economic development is essentially based on the exploitation of natural resources and environmental pollution, which becomes the subject of a separate study. Modern business shows a particularly strong connection between energy consumption and development in general. The modern world consumes large amounts of energy, and the intensity of energy production and consumption is different in different parts of the world. Energy production and consumption have a direct impact on environmental quality, but they are a necessary precondition for economic progress. To determine the balance between the consequences of energy production and consumption, on the one hand, and welfare resulting from economic development, on the other, is a very sensitive issue. There are numerous efforts to create a balance between energy resources, consumption, and satisfactory development, which are linked with a great number of global problems and disagreements on all levels (Iddrisu & Bhattacharyya, 2015).

When planning development, each company should take into account all the consequences that will happen after introducing particular changes. A sustainable approach to business and energy management presents a big change that brings about certain consequences, whereby the changes in the economic sphere are of special importance. While the sustainable approach is desirable, it does not lead to economic collapse of the company. Thus, it is necessary to consider and plan the company development with respect to the rules of sustainability, but that should not threaten a company's survival.

Globally, sustainable energy management has an economic effect that can be viewed on two grounds (Park & Kwon, 2017):

- Sustainable energy management in a company produces a range of positive effects, whereby positive economic effects arise primarily as a result of savings arising from the proper implementation of sustainable energy management; and
- Realization of projects from sustainable energy management and exploitation of alternative energy sources are a means to achieve certain incomes, where these investments are treated as separate business ventures.

Both these approaches offer the possibility of considering sustainable energy management as a change in the company conditioned by the need to operate in a socially responsible manner while also having strong economic justification.

These basic characteristics of energy production from alternative sources point to the fact that obtaining energy in this way is certainly desirable from the aspect of sustainable energy management and should be considered in each situation.

These amounts must be considered along with other indicators in each country. Government subsidies given to producers of alternative energy should also be taken into account. In many countries (the EU region and all other European countries), producers of alternative energy have a privileged status, they receive certain subsidies, and the state is obliged to purchase all the energy they produce as a preemptive right (Özdemir, Hobbs, Van Hout, & Koutstaal, 2020).

Second strategic objective: Energy efficiency

Energy efficiency

Definition

Defining energy efficiency involves consideration of several different approaches but covers a wide range of energy consumption indicators.

Energy efficiency encompasses all changes that result in a reduction in the energy used for a given energy service or level of activity. This reduction in energy consumption is not necessarily associated with technical changes, since it can also result from better organization and management or improved economic efficiency in the sector (e.g., overall gains of productivity). This definition of energy efficiency is provided by the World Energy Council (WEC) (Pérez-Lombard, Ortiz, & Velázquez, 2012).

Energy efficiency improvements refer to a reduction in the energy used for a given service or level of activity. The reduction in energy consumption is usually associated with technological changes, but not always since it can also result from better organization and management or improved economic conditions in the sector ("nontechnical factors").

In some cases, because of financial constraints imposed by high energy prices, consumers may decrease their energy consumption through a reduction in their energy services (e.g., reduction of home temperature or car mileage). Such reductions do not necessarily result in increased overall energy efficiency of the economy and are easily reversible. They should not be associated with energy efficiency.

To economists, energy efficiency has a broader meaning: it encompasses all changes that result in decreasing the amount of energy used to produce one unit of economic activity (e.g., the energy used per unit of GDP or value added). Energy efficiency is associated with economic efficiency and includes technological, behavioral, and economic changes.

Energy efficiency is first a matter of individual behavior and reflects the rationale of energy consumers. Avoiding unnecessary consumption of energy or choosing the most appropriate equipment to reduce the cost of energy helps to decrease individual energy consumption without decreasing individual welfare.

Avoiding unnecessary consumption is certainly a matter of individual behavior, but it is also often a matter of appropriate equipment: thermal regulation of room temperature and automatic deactivation of lights in unoccupied hotel rooms serve as good examples of how equipment can reduce the influence of individual behavior.

Energy efficiency is a concept that differs from the concept of energy management. Energy efficiency essentially means consumption of less energy to obtain the same level of energy service, whereby energy management means the planning, control, and optimization of the energy use, in particular toward achieving energy goals.

The main elements of an energy efficiency monitoring process

Energy efficiency monitoring is in essence a very complex process, consisting of several elements (input, output, and activities).

Inputs that characterize energy efficiency can be defined by many criteria separately and/or simultaneously. In addition, the measurement can be performed using conventional or modern, more complex methods of measurement, using various metrics (Castro Camioto, Moralles, Mariano, & Nascimento Rebelatto, 2016). The energy is obtained from various sources in various forms, the distribution and transformation of energy from one form are frequent and linked with losses, and contemporary needs are based on the use of energy in different forms and quantities.

Defining the output elements of an energy efficiency process is also complex and subject to a great number of factors and needs. In the majority of cases, output indicators, used to determine the level of energy efficiency, are determined by the energy that is transformed and incorporated into the final product or service, which is added to the waste energy and energy that is lost in the form of externalities. Definition of waste energy and the energy lost in the form of externalities is linked with many problems. The methodology for its determination is undefined and incomplete, and the need for more precise work in this area is evident. Only a precise determination of the total output power can ultimately determine the level of energy efficiency and its changes (Freire-González & Vivanco, 2017).

All activities that occur in the process of changing energy efficiency are the subject of particular interest and constant improvement. The improvement of energy efficiency is an activity defined as one of the strategic objectives of improvement in the energy management process, both on global and national levels and at the level of each company individually. Changes in the level of energy efficiency may result from the implementation of various activities (of technical and/or organizational nature), and opportunities for improvement are found in virtually every sphere of activity associated with energy consumption.

The main principles of global energy efficiency policy

Reconciliation of a company's plan with global and national energy efficiency policies means that the company's policy will be compiled in a way that will be fully adjusted to the basic regulatory documents existing in the particular area of the country, region, and world. Globality of the energy-related problem has created the need for cooperation at all levels, and compliance with global policies in this field is imposed as one of the imperatives. All strategic documents existing at the global level recognize the problem of energy efficiency as a key one, and accordingly, resolving and improving are placed on the list of long-term priorities (in the energy policies of the United States, Canada, Japan, European Union, and Russia). In all these countries and regions, the improvement of energy efficiency, along with the increase in the volume of energy generated from renewable sources and reduction of GHG emissions, is one of the top three priorities (Inshankov, Bogachkova, & Popkova, 2018).

A special aspect of global energy efficiency policy includes the development of clean technologies that will contribute to the implementation of energy efficiency improvement. There are specific standards and methods that support and promote energy-efficient technologies and products. Energy efficiency is being studied, and the solutions for use in individual households are being improved. In addition to technological improvements, the mechanisms and instruments that accompany the economic aspects of energy efficiency are being developed, implementing changes in the way of determining GDP and other conventional methods for determining the economic indicators of the condition.

One of the goals of global energy policy is to increase the amount of energy obtained from renewable sources. The US and EU strategic documents define that the amount of energy produced from renewable sources shall be 20% by 2020. The technologies of energy production from renewable sources are being improved. There are special procedures that support investing in this area.

Reduction of GHG emissions and other emissions occurring in the production, distribution, and consumption of energy (primarily CO_2, NO_2, and SO_x) is also defined as a strategic objective at the global level. Technological solutions that contribute to reducing these emissions, as well as methods of measurement, monitoring, and trading carbon credits, are being developed.

Energy efficiency management is a process based on several assumptions. First of all, it is designed in a way to ultimately enable improvement of energy efficiency parameters both in physical and in economic terms. In addition, the energy efficiency management process must be adequately integrated into all other management processes within the company as well as the process of strategic management in general (Trianni, Cagno, & Farné, 2016).

Management in this area is aligned to the general starting points that observe the problem of increasing energy efficiency as one of the priorities of sustainable development in general. Therefore, any change in a positive direction can be considered acceptable and fully justified. The management process must be designed to ensure positive progress, no matter how small. Management based on monitoring incremental changes is imperative for energy efficiency management.

The energy efficiency management process can be viewed in two ways. On the one hand, it can be seen as part of a sustainable energy management process and, on the other hand, as an independent process. Sustainable energy management is a complex process based on implementation of management processes that comply with specific production and consumption, and goals are aligned with the basic objectives of sustainable development. Both observation methods are considered necessary. Energy efficiency is just one of the parameters in the complex process of sustainable energy management, and it must be regarded as one of its main goals. Energy efficiency is the goal, and energy management is a way to achieve this goal.

Energy efficiency audits

The importance of energy efficiency audits

Modern society, which is turning to operations based on the principles of sustainability, largely puts the problem of energy consumption at the forefront. Development plans incorporate issues associated with energy into strategic objectives and undertake extensive activities that enable the improvement of the most important indicators of sustainable energy development, such as reduction of GHG emissions, increase in the energy production from renewable sources, and increase in energy efficiency.

Given the nature of activities, the increase in energy efficiency can be regarded as potentially the most important indicator of success of the company in the contemporary and future long-term period. Energy efficiency is the amount of energy that is used for obtaining a certain number of products and is expressed in physical size. Much greater importance is attached to the indicator of energy efficiency as measure of the amount of money spent to cover energy costs for performing these activities. Energy efficiency as an economic size is a particularly important indicator in the following cases (Reuter, Patel, Eichhammer, Lapillonne, & Pollier, 2020):

- Companies that perform activities that are, by their nature, great energy consumers, which in most of the cases means energy generated from nonrenewable sources. Globally, most companies today consume energy in a quantity equal to or greater than the amount that the company spent in the previous period. In the developed world, companies pay increasing attention to reducing energy consumption, but generally speaking the average consumption is still high;
- Companies spending imported energy-generating products, which greatly increase the level of import dependence and threaten the security of both the company and the country in which it operates;
- Companies that use outdated technology are particularly large consumers of energy, and they exist in most of the world. Only a small number of countries are in a position to make significant investment in installing modern energy-friendly technologies. Regardless of the extensive evidence of profitability of the investment after a certain period of time, most companies in the world are not able to invest large sums of money in purchasing new technology. By exploiting old technologies, which are, besides high-energy consumers, prone to disorder, the company uses more energy, and in that manner, it further invests in the losses, thereby reducing the amount of income that is available for investment; and
- Companies that are highly environmentally aware also see energy efficiency audits as an activity that provides them with certain required information. In this way, they can plan future activities and adjust future performance for improving energy efficiency and sustainable business in every sense.

Energy efficiency audits provide companies with data so they can get an accurate picture of the degree of sustainable business in the area of energy consumption, the environmental load, which thereby occurs (primarily GHG emissions), and economic effects that occur at the same time. Measuring and monitoring energy consumption using only quantitative measures are not considered sufficient for the purpose of obtaining complete information. It is necessary to monitor indicators of energy consumption per unit of activity per certain working step, process, and unit of product. The full picture is obtained in the end by measuring indicators of external costs, which represent the economic value of environmental pollution caused by energy exploitation (Kabule, Ločmelis, & Blumberga, 2020).

Energy efficiency audit objectives

Depending on the type of activity, size of company, availability of energy resources, and other parameters, different objectives for performing energy efficiency audits can be defined, but in most cases, the objectives include (Kluczek & Olszewski, 2017):

- Identifying the type, quality, quantity, and price of energy resources used for activities of the company;
- Special monitoring of energy consumption from renewable sources;
- Definition of consumption units (profit centers—work processes, operations, technological lines) on which the monitoring of energy efficiency indicators will be performed;
- Determining the energy cost of previously defined profit centers, thus providing insight into the energy efficiency of individual profit centers;
- Defining all parameters of energy inputs and outputs;
- Identifying internal energy transportation routes and potential losses;
- Identifying the place of origin, quantity, and characteristics of energy waste, along with special monitoring of pollution that is the subject of legislation and internationally accepted agreements;
- Monitoring all of the previously listed indicators by hours, days, months, seasons, and years, identifying regularities and irregularities in certain form, and defining the cause; and
- Determining savings in some profit centers, whereby the potential savings, investment required, the list of priorities, and expected results are particularly defined.

An energy efficiency audit, through its objectives, obtains large amounts of data, which lead to one, final result. The main objective of an energy efficiency audit is to get a broader picture of the level of sustainable energy company operations, including energy, environmental, and economic indicators. In this way, companies can assess their current position and plan future activities.

Energy efficiency audit levels

Level I audit

The Level I energy audit includes an overview of business' annual energy consumption and a walkthrough to evaluate the performance of existing facilities. This audit will encourage businesses to reduce energy use by 10%. It includes the inspection of a facility to identify specific issues and needs. The audit provides a description of the facility, including (Andersson, Arfwidsson, & Thollander, 2018):

- Brief description of what type of activity occurs in the facility;
- Square footage;
- Year built;
- Building envelope and insulation;
- A summary of business' annual gas consumption by month;
- Operation and maintenance recommendations; and
- A brief overview of equipment replacement and projected annual energy savings.

Level II audit

The second phase is a more in-depth study that includes capital cost items such as boiler or burner replacements, energy management upgrades, etc. If applicable, this audit will work in cooperation with an electric utility. The Level II audit provides a comprehensive report with detailed energy efficiency recommendations. This report includes the following (Locmelis, Blumberga, Blumberga, & Kubule, 2020):

- On-site inspection of the facility to identify specific issues and needs, including
- building envelope, windows/doors, sidewall and attic, and insulation levels;
- Inspection of HVAC equipment and current operation of this equipment;
- Evaluation of HVAC system efficiency;
- Operations and maintenance procedures;
- A description of the facility, including description of what type of activity occurs in the facility, square footage, year built, building envelope and insulation, HVAC system, domestic hot water, commercial cooking, and age;
- Major equipment list including hours of operation;
- A summary of important findings along with any supporting data, measurements, drawings, etc.;
- A summary of business' annual gas consumption by month;
- Gas usage pie chart with calculated annual heating costs for windows/ doors, attic/roof, sidewalls, fresh air ventilation including domestic hot water, laundry, kitchen appliances, and other major equipment;

operation and maintenance recommendations, including ways to optimize equipment operation, update energy management protocol, and improve maintenance practices and operational changes;
- Major equipment investment recommendations, including
- replacement of HVAC equipment size calculation based upon heat loss and ventilation analysis;
- Calculated annual energy savings based upon building space; and
- Economic analysis (payback, ROI, etc.).

The proposed model of energy efficiency audit is based on the average needs of the organization and should be changed and adjusted for specific needs. However, all audit techniques should simultaneously observe both physical and economic indicators of energy flow and thus provide information relating to the energy and economic viability of the company.

Improvement of energy efficiency

The changes in energy prices are the best and a natural driver of activities aimed at increasing energy efficiency in all sectors and their establishment, along with the necessary social programs of the state for those categories of energy users that need to be subsidized.

The practice of developed countries has shown that the state must play a leading role in systemic actions aimed at increasing energy efficiency. This is especially necessary in economies where the cost of energy and energy-generated products is controlled by the state. In such conditions, end users see energy as a necessary cost, the reduction of which, as a rule, is not a priority. This whole area of activity aimed at increasing energy efficiency discourages interest for investing in such projects. When energy and energy-generated products become goods fully subject to all rules of the supply and demand law, the leading role of the state will no longer be necessary. Free investment funds due to their nature "look for" projects where the realization will increase their value, by increasing energy efficiency (Du, Matisoff, Wang, & Liu, 2016).

Until these barriers are removed, the leading role of the state in terms of increasing energy efficiency is reflected in the creation of adequate legal framework (laws, bylaws, regulations, etc.), in creating incentives for the implementation of a number of technical measures that lead to its increase, and in influencing the behavior of end users by raising their awareness in this area.

In addition, the establishment of liability in respect to energy management (introduction of energy management) and a constant concern about increasing energy efficiency in all sectors of energy consumption and production are a prerequisite for the implementation of systematic measures to increase energy efficiency in the country as a whole.

Bylaws govern the rational use of energy, among other things, so it is important to regulate the obligation of energy entities (energy users and consumers) to establish appropriate units for management of energy, so-called energy management, which should follow energy parameters and periodically report to the relevant government institutions on the current parameters of energy, promote energy efficiency by taking measures for its increase, and report to relevant government institutions achieved results, observing the prescribed procedure.

Encouraging energy efficiency can be enhanced by applying specific measures, and the following can be considered the most important (Tallini & Cedola, 2016):

- Providing a variety of tax and other incentives for companies that implement projects in order to improve energy efficiency;
- Introducing the principles of energy efficiency in public procurement and in the criteria for the allocation of other funds and loans;
- Ability to support projects aimed at increasing energy efficiency;
- Cooperation with international financial institutions in order to realize favorable credit arrangements;
- Promoting the interest and building capacity of local banks to lend to projects in the field of rational use of energy;
- Creation of legal conditions for a special type of financing by "third parties" or for the operations of companies that implement saving measures (in the companies or municipalities) and pay their for services from the achieved energy savings; and
- Ratification of the Kyoto Protocol and the implementation of other necessary activities in order to open opportunities for the use of Clean Development Mechanism under the Kyoto Protocol for the realization of energy efficiency projects. The Clean Development Mechanism (CDM) is one of the flexibility mechanisms defined in the Kyoto Protocol. It is defined in Article 12 of the Protocol and is intended to meet two objectives: (1) to assist parties not included in Annex I in achieving sustainable development and in contributing to the ultimate objective of the United Nations Framework Convention on Climate Change (UNFCCC), which is to prevent dangerous climate change; and (2) to assist parties included in Annex I in achieving compliance with their quantified emission limitation and reduction commitments (GHG emission caps).

In addition to these measures, it is also necessary to work on raising awareness of end users, the implementation of education in this area, and the implementation of demonstration projects to identify and promote good practices.

Technical and organizational measures

The basic organizational measure aimed at the rational use of energies implies the introduction of managing energy (energy management) for all consumers whose total installed capacity is more than 1 MW. This measure involves the monitoring of energy consumption and constant concern about increasing energy efficiency.

As for the sectors of industry and building construction, energy audits aimed at recording the current status of spending and identification of measures to improve energy efficiency is of great importance. In this sector of industry, a 15% reduction of energy consumption may be achieved between 15% and 20% (Hrovatin, Dolšak, & Zorić, 2016).

Significant potential is found in improving the combustion process along with regular auditing of this process. This measure can reduce energy consumption by 2%–3%, equal to 0.0809 million ton (0.9402 TWh) or 28.2 million euro decrease in fuel costs. For the implementation of these measures, additional funds are not required, since existing maintenance funds can be used. However, it is necessary to adopt a regulation and for energy entities to implement it, whereby the inspection will strictly control the execution of this obligation.

By adding condensing areas on boilers in industry and in heating plants designed and built for heavy oil where natural gas is currently used, an increase in the energy efficiency of the boiler units of 6%–7% can be achieved. This investment should be considered since as a rule, it is repaid in less than 1 year if the boiler works 4500 h/year. This measure can be achieved through partial stimulus by the country or through the company's own funds.

Using waste heat from power plants and production processes has the potential to increase energy efficiency that can reach up to 20% of the total heat production needs (Broin, Nässén, & Johnsson, 2015). This is typical for the drying, processes in the chemical and food industry, industry of building materials, and the like.

The energy integration of the production process, especially in the chemical industry, presents a potential that can be used to increase energy efficiency of a factory's heating systems up to 5% with a relatively short repayment period (usually less than 1 year and not longer than 3 years).

Replacement of the existing engines. This measure can be explained in the following example. 4000 MW of electric motors were installed. Replacing them with engines having greater efficiency of class EF1 and EF2 reduces consumption of electricity of 188 GWh/year or 7.5 million euro. Investment for the implementation of these measures is 200 million euro. This measure provides substantial energy effects, but given the current electricity prices, its implementation is economically unrealistic, since due to the present conditions of factories and parity pricing to end users, it brings effects only after 15 years.

Transport sector

Measures to be implemented in the transport sector include the following (Feng & Wang, 2018):

- Defining, identifying, and adopting a national strategy on the development of transport systems. The national strategy on the development of transport is necessary, since the current situation is alarming in all sectors.
- Rejuvenation of the fleet in all sectors. Besides the other issues and considering the energy efficiency aspect, the age of the fleet is among the key issues. It is therefore necessary to adopt measures to stimulate the purchase of new cars and discourage the use of vehicles older than 15 years. Each country has its own regulations regarding car import.

Building sector

The following activities have been defined in this sector (Alla, Bianco, Tagliafico, & Scarpa, 2020):

- Switching from heating using electrical energy to other forms of energy. It is estimated that it is possible to achieve a reduction of the current electric power system load for heating using electrical energy, which currently amounts to 3062 MW, by 1222 MW by replacing it with the systems of remote and central heating or natural gas, which would result in annual savings in electrical energy of 1500 GWh. The value of that energy is, at the current purchase price of electricity in the market, 60 million euros a year. This requires a planned construction of these systems in areas dominated by heating with electricity. It is likely that these zones are suitable for the development of natural gas for easier construction and less expensive installation. However, due to the possible amount of investment, this measure requires detailed individual analysis of each site separately.

- Replacement of lightbulbs in homes and commercial buildings. By replacing only two lightbulbs of 100 W with corresponding fluorescents of 20 W, an electricity savings of 701 GWh can be achieved and the source capacity of 480 MW released, i.e., 28 million euros. Investing in the implementation of this measure (purchase of bulbs) amounts to 18 million euros with a repayment period of 8 months. Implementation of these measures is possible in various ways.
- Adoption of new regulations on the outdoor design temperature. The aim of this measure is the reduction of the projected installed power of home installations and heating sources for buildings, improving energy efficiency in the use of heating systems, and reduction of capital costs for these plants. Reduction of the middle design temperature at the level of the state by 40°C would result in 10% energy savings for heating (the implementation of this measure would eliminate the currently evident heating above the projected values). Implementation of the measure does not require investment funds.
- Consistent application of ISO 14000 and other standards. Consistent application of basic and other related standards for the design of new buildings and their thermal protection can reduce the design installed capacity of heating by 30%–40% and achieve approximately such energy savings for heating. Implementation of measures requires only strict enforcement and compliance with existing regulations, as well as the achievement of real price of apartments with the energy quality control of the facility on the handover of the buildings and heating systems in them. It is necessary to adopt a maximum allowed value of the final energy for heating buildings of kWh/m^2 year and to expect the designer to meet the stated value in terms of heating physical characteristics. Data on annual consumption of energy should be an integral part of the documentation of each building. Licensed individuals would be expected to determine consumption based on the project.
- Establishment of incentive funds aimed at improving the thermal protection of existing residential buildings. It is necessary to establish incentive funds for improving the thermal protection of existing residential and nonresidential buildings at national, province, and local levels. For that purpose, the budget, dedicated foreign and domestic credit, and donor funds are to be used, with the mandatory participation of owners in cofinancing of additional thermal protection of buildings.
- To replace the windows in all multifamily residential buildings connected to central heating systems, in 60% of multifamily residential

buildings that are now using electricity in all single family buildings with central heating, and 40% of single family residential buildings using electricity or natural gas for heating, the amount of 936 million euros would be necessary for multifamily residential buildings, or 1326 million euros for single residential buildings. The overall effect of the final energy savings for heating upon the realization of this investment amounts to about 7.4 TWh/year (4.05 TWh for multifamily and 3.35 TWh for single family residential buildings), and if the money is invested now, the total repayment period would amount to 8.9 years (for multifamily 7.1 years with an annual saving of 132 million euros and for single family 10.8 years with annual savings of 122.7 million euros).

- For the purpose of improving the insulation of the walls of the facilities, it is necessary to invest 341 million euros for multifamily residential buildings, and 497.9 million euros for single-family residential buildings. The total effect of the realization of energy saving activities indicated above is the savings in the final heating of about 3.5 TWh/year (1.9 TWh for multifamily residential buildings and 1.6 TWh for single family residential buildings). In the period from 2007 to 2012, it is envisioned that the investments would achieve 10% savings of the stated potential.
- Adoption of regulations for the constructors of new buildings and owners of the existing building to obtain a certificate of energy efficiency for each building.
- For new buildings, which will be heated from the system of remote or central heating, it is necessary to use TPV in the substations and boilers of these systems.

Only basic measures to increase long-term energy efficiency have been proposed. Given the fact that saving of energy and its rational use is one of the imperatives of sustainable energy management, determination of new measures, specific to a particular activity, is under development.

Third strategic objective: Risk management

Energy management risks

Energy management is a complex, long-lasting process conditioned by numerous different factors. As such, it is implemented in an environment that is more or less unstable, where the environment influencing the processes of getting, distributing, and consuming energy in the world is particularly variable and unpredictable. Certainly, there exist and are applied numerous possibilities to predict fluctuation of the global energy market, but particularities and significance of energy for overall development inevitably led to problems and events that are difficult to manage. Energy management represents the process attached to increased risk and thus must be planned and implemented with special care.

Sustainable energy management is a layered energy management process, and thus it becomes an even more complex process with the risk assessment that accompanies and influences it. Risk, as a probability of a specific event, represents a complex phenomenon that can be studied and observed from many aspects. In the process of sustainable energy management, risk can be regarded as an individual process that is integrated with the sustainable energy management process (Ioannou, Angus, & Brennan, 2017).

During risk assessment in the process of sustainable energy management, it is necessary to take into account and analyze numerous internal and external factors. Apart from that, it is necessary to analyze economic and noneconomic risk factors that can influence the success of sustainable energy management implementation in a particular business system.

Economic risk factors include a large group of events that can occur in an economic environment, which above all includes (Petrović, Filipović, & Radovanović, 2018):

- Supply market—where it is necessary to estimate all factors of energy source supply, their availability, purchase price, distribution terms, quality, and specificities with specific analysis of energy supply capacity produced by alternative resources. It is also necessary to analyze capacity to produce energy from domestic resources;
- Sales market—a company that decides to implement process of sustainable energy management at the market discloses its product enhanced with the quality in the form of responsible behavior, and as such it must find its place in the range of all other products on the market. In certain cases, products of such companies may cost a bit more since introducing the process of sustainable energy management demands high initial investment. Thus, it is necessary to analyze market competence to react positively to emersion of such products;
- Competition—every company deals with existing competition. It is necessary to consider the strengths and quality of the competition, its geographic distribution, and the market segments that it covers. Contemporary companies, among other things, fight for their customers by practicing environmentally responsible business, so that gaining strong competitive position in great measure depends on behavior toward the use of energy resources;

- Legislation—every country has certain law regulations defining business regulations, and every change can result in major problems for the company. Modern legislation in great measure supports sustainable energy management, but risks of changes in this sphere are still possible;
- Technology innovations—are of particular importance in the process of sustainable energy development. All changes in technology must be specifically anticipated. It is also necessary to take into account quality, price, availability, delivery, and installation of certain technology as well as other factors of this kind; and
- Quality product regulation—modern quality parameters in great measure demand products of particular quality regarding the energy exploited for their production, as well as energy characteristics of products where energy saving and energy efficiency are set as top priority.

Noneconomic factors are a sequence of various factors that in great measure can have positive or negative influence on the implementing process of sustainable energy management. Regarding the specificity of global distribution and significance of energy sources for human race development, it is necessary to analyze these factors as precisely as possible (Filipović, Radovanović, & Golušin, 2018):

- Availability of energy resources—uneven distribution of energy sources puts a great number of world economies into unequal positions, since many countries are bound to import energy sources. Thus, every country strives to provide a sufficient amount of energy for its own functioning, but this greatly depends on the regulations dictated by the countries disposing of energy resources;
- Conditions of natural disaster occurrence—energy resources are the products of countries that are greatly vulnerable to the possibility of a natural disaster, which can greatly endanger their capacity to supply the world market;
- Conditions of disaster occurrence due to human activities—facilities for producing and processing energy, as well as objects and means of distribution, are in their essence complex, and as such they can be particularly subject to occurrence of minor or larger industrial disasters;
- Political factors—are the factors that determine functioning of the world today in great measure, especially in the energy distribution sphere, which is a limited resource. All political changes must be considered, and special attention should be given to political predictions in order to minimize the risk caused by their occurrence; and

- Warfare—risk of military action is always present in regions disposing of large energy source supplies, which represents continuous risk for energy source availability to all, depending on energy import.

These are only fundamental economic and noneconomic factors that need to be taken into account when estimating the risk of performing a sustainable energy management process. Their analysis is complex but necessary since transition from the ways of traditional to sustainable energy management represents a kind of strategic change that inevitably leads toward big and long-lasting changes.

Basic steps of risk management

Risk assessment includes performing whole sequences of assessments that are extremely complex for activities that are associated with energy production and consumption since they might lead to environmental accidents with huge consequences. Facilities dealing with energy production and transport in a conventional way are the ones carrying the high environmental risk, while obtaining energy from renewable sources may be regarded as environmentally quite acceptable. Obtaining energy from nuclear power plants represents a particular problem in risk assessment. The risk management process consists of three basic steps, as shown in Fig. 7.23.

Identifying the risk is the first step in the process of estimating the risk for the environment. Estimating the risk means observing company technology process from the aspect of environmental protection. When observing technology this way, it is most important to realistically examine all ongoing events (Fig. 7.24).

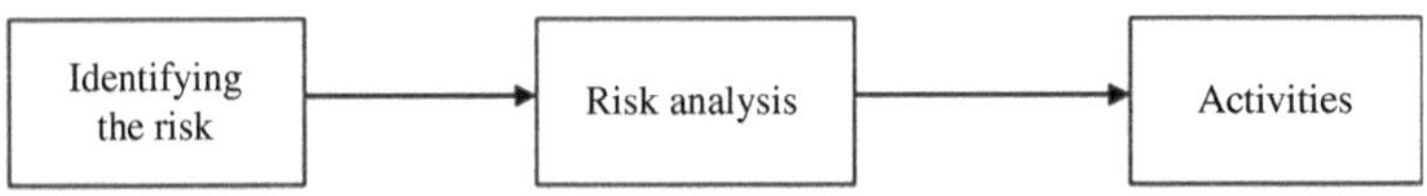

Fig. 7.23 Risk management process.

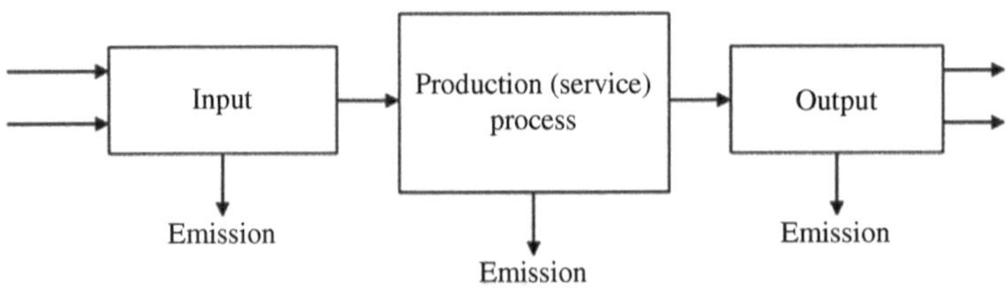

Fig. 7.24 Basic steps in identifying the risk.

When identifying the risk, it is necessary to analyze the production process in detail and to note all pollution points that arise as a consequence of energy consumption that represent potential risks. It is quite possible that during any phase of the production process, hazardous emissions are affecting the environment that can be easily and quickly eliminated with a minor investment of money and labor, but there are also events that are difficult to manage. When identifying risk points, it is priority to note events that are highly dangerous that primarily affect closest work environment. Modern technologies are constructed so that a number of risk points are minimized as much as possible, but risk estimating process remains the necessary part of the risk management.

Risk analysis, as a part of the risk management process, implies analyzing every observed critical point. In the first phase, it is necessary to note all parameters of observed risk, including:

- Risk origin, where it should be determined why the critical point in the technology process has occurred in a specific place, i.e., what caused the pollution. Proper evaluation of pollution origin is of particular significance since it can provide valuable information that can instantly offer adequate solutions. It is a different situation if it turns out that the cause of the harmful effect is a human factor, faulty installation of the facility, a bad technological solution, etc.;
- Type of pollution means detecting pollution type, i.e., its physical, chemical, and biological nature;
- Activity direction, the various waste substances directed toward different parts of nature; some pollute the air, some water, some directly affect health of people;
- Effect intensity, defining the strength of the polluting emission effect; and
- Hazard, the final analysis, giving the real danger of pollution to the environment.

Upon evaluating characteristics of the risk event itself, a list of priorities can be determined. Energy consumption is often associated with risks of various intensities, but in its essence, it is the cause of a great number of risks that exist in performing certain activities. The energy process is the process expected to cause pollution and thus requires particular attention.

Risk evaluation, where all previously determined parameters are perceived together and a decision (evaluation) is made where individual critical points are labeled as a high, medium, or low risk. Further actions depend on this evaluation. High-risk influences should be eliminated as soon as possible, while low-risk influences in most cases can be eliminated in a simple way

(Jackson, 2010). Creating a priority list is important in every management process since it is inefficient and unproductive to invest efforts in eliminating all observed harmful influences emerging due to energy consumption in different places within the company. Experience shows that goals must be ranked according to urgency level and solution possibility, and afterward they should be approached individually. Only when one problem caused by energy consumption is solved can the second problem can be approached.

Activities include sequences of activities (applying protection measures, employee training, strict and continuous control) performed in order to adequately detect critical points by previous procedures. As an example, it is nuclear power plants or industry with particularly dangerous chemical compounds carrying high environmental risk, but risks are controlled and nothing is left to chance.

It is possible to establish that a certain technology has a great number of critical points whose rehabilitation is quite expensive and thus unprofitable. In such cases it is indicative to consider changing the technological process or changing the whole technology. As a rule, technologies are energy-efficient and energy consumption is controlled. Nevertheless, modern technologies are not available to all in the world, and the problem of transport is still present.

Finally, it is necessary to emphasize that environmental risk assessment is a necessary parameter when evaluating future investment activities, i.e., making decisions about beginning new activities. Every business or investment project must provide information on the possible environmental risk that a new business carries as well as provide information on the method by which the company will keep these risks under control.

Risk assessment

Risk is an event that can be of a different nature or intensity. Risk assessment is in its essence a complex and multifactor analysis associated with all business aspects and is particularly analyzed for the purpose of business economic adequacy assessment. In fact, certain risks may bring into question a company's survival or demand investments that are so large they impose strategic shifts in doing business with far-reaching consequences. Modern business rules and the concept of sustainable development demand operation risk assessment regarding energy consumption of existing companies as well as of investment analysis, i.e., investment in new activities. There are three fundamental questions that should be considered prior to discussing the ranking of the project and difficulty criteria for managing project finance (Taroun, 2014).

First, it is necessary to estimate the strategic scope of the analyzed project or venture, i.e., to estimate to what extent the project should be strategic and opportunistic. Projects with distinct strategic character should be subjected to detailed analysis. Most projects relating to energy consumption can be regarded as projects of this type since the energy is a strategic resource and every activity in this field is of particular significance.

Second, it is necessary to analyze existing types of risk. In conventional portfolio management it is understood that all investments, regardless of their potential profitability, may carry significant failure risk. This is especially true in the case of an environmental investment whose potential profitability may depend on uncontrolled risks unevenly distributed in the observed region.

Third, it is necessary to take into account the economic implications the project entails. Regardless of the priority, sustainable energy development has methods to achieve equity and equivalence by balancing means according to the demand of achieving the biggest environmental profit per money unit spent and should be explored. It is necessary to estimate whether indicators should be set so that they justify forced projects in order to achieve goals that are not economy oriented. Energy production and energy resource management represent an issue of strategic importance to every country, so projects of this kind cannot be regarded solely as economic ventures.

Assessment of potential risk that a certain project entails is extremely complex, and the assessment of the energy risk makes it even more complex. Global problems boosting the risk in this sphere further complicate the problem. Namely, problems of uneven distribution of energy resources, conditioning the supply of energy resources, unstable markets, and price fluctuations are only some of the risk factors that should be taken into account. General factors that should be considered include:

- Exposure—it is necessary to estimate how many and what kind of uncontrolled risk factors may influence potential results. Consultants' help can be quite efficiently used for exposure analysis;
- Timelines—it is necessary to examine whether there is a trend where major risk factors become more or less favorable. If the assessment of time is wrong, this can lead to project failure due to premature investment, or it can even become profitable after a long time. Most requests from the sphere of sustainable energy management are strongly

expressed, and it is expected that in the future this will be even more true. Thus, it can be considered that the timeline factor for energy management will have a stimulating influence on the projects of this sphere;

- Volatility—it is necessary to estimate the probability of the anticipated result if all uncontrolled risk factors act together. In the case of analyzing this factor, it is really important to have a good plan for the investment since this factor can have a devastating impact if all kinds of risk factors occur simultaneously or within a short period. For projects dealing with energy, a high level of these influences can be expected, which is a consequence of instability in the global energy market, and thus this analysis should be performed carefully;
- Safety—regardless of the number of risks, it is necessary to consider whether there is a way to minimize or to decrease the influence of uncontrolled risk factors. This mostly refers to the phase in the risk management process where the reaction decision is made. Safety is mostly achieved by risk transfer or by insuring the project outcome; and
- Persistence—projects in the energy industry are mostly long-term projects, so it is necessary to estimate how long the uncontrolled risk factors will influence the success of project. In this way the possibility of activity termination of the noted risks during the process can be perceived. It is also necessary to predict whether the activity of uncontrolled risk factors will continue after the project is finished.

The above includes only the fundamental factors that should be taken into account when planning the investment necessary for the implementation of sustainable energy management. In case that process predicts major investments (which often is the case), risk analysis must be performed precisely. Investments in the process of sustainable energy management are quite often initiated as a result of a company's orientation toward socially acceptable business, so projects are long-lasting, demand major initial investment, and positive economic effects take place a few years later (Chebotareva, Strielkowski, & Streimikiene, 2020). Because of this, the process of investing in sustainable energy management, although environmentally justified, must be performed in a way that will not endanger a company's long-term survival.

Methods of environmental investments risk assessment

In practice, there are two basic groups of methods to estimate and measure environmental investment risk: nonmonetary and monetary methods.

Depending on the need, these methods are used in order to perform risk assessment to provide information about the extent to which the risk of investing in sustainable energy management is a risky venture, achieved by risk quantification (Liebe, Gewinner, & Diekmann, 2018).

Nonmonetary methods are methods that are in the development phase and are featured by specific advantages and disadvantages. The main advantages of nonmonetary methods are that they can focus on necessary terms to provide services, can be based on conventional tools, reflect factors that determine aggregate payment capacity, reflect risks of service process failure, can be used as a benefit-transfer method, can be used to create lists of environmental priorities, and can reflect correspondent preferences for the product mix. The main disadvantages are that they cannot be used in conventional cost-benefit analysis, they are not particularly useful to justify increase in costs, are generally not popular with economists and scientists, demand certain decisions that must be additionally estimated, and demand initial investing in data acquisition; and.

Monetary methods are methods based on the assessment of the currency value and thus have certain advantages and disadvantages. The main advantages of risk assessment based on the monetary method are that they are a generally accepted economic benefit measurement method, are easy to interpret and use for the purpose of comparison, practical to use in conventional cost-benefit analysis, and for certain individuals they are the only acceptable measurement method. The main disadvantages of monetary methods lie in the fact that evaluation methods of nonmarket products are not sufficiently developed, and they are expensive to implement. Results of nonmarket evaluations are often conservative and in most cases specific for a certain region. There are difficulties in determining the connection between benefits and decision and utility difficulties in order to create environment priority list. Monetary methods are mostly used for investment risk assessments in investments with sustainable energy management as a goal. Basic monetary methods have been developed for practical applications:

Market price method—this method estimates the economic value of the ecosystem product or service that can be bought or sold on the market. This method can be used to estimate changes in quality and quantity of the goods and services. It uses standard economic techniques to measure the economic benefit of market goods based on quality of the supply by different prices. The standard method for the measurement of utility values that can be found on the market is the assessment of consumer and producer surplus. The economic benefit total or economic surplus represents the sum of consumer and producer surplus. Assessment of the energy amount used for obtaining each product is in the essence of this method.

Productivity method—this method estimates the economic value of the ecosystem product or service that supports production of commercial market goods. The productivity method is also regarded as the derived value method used to estimate the economic values of an ecosystem product or service involved in production of market goods and services. It can be applied when products or services are used together with other inputs for the production of market goods. For example, the use of alternative energy resources in some production processes improves the value of the final product and thus becomes environmentally acceptable. The economic benefit obtained by the production of the energy from alternative resources can be measured by the increase of revenue resulting from the consumption of energy from traditional sources.

Hedonic pricing method—this method estimates the economic value of environmental services that directly influence the market price of other goods. This method is used for the assessment of an economic value of environmental service. It is mostly applied to determining house prices that also reflect the value of environmental attributes. It takes into account environmental qualities (air pollution, water pollution, noise, etc.) as well as environmental comforts (nice view, aesthetic values, etc.). The basic premise of the hedonic pricing method lies in the fact that the price of market goods is connected to their characteristics or to the services they provide. For example, the price of a car reflects some of its characteristics—its ride, comfort, luxury, cost effectiveness, and so on. Thus we can evaluate certain characteristics of a car or some other market good based on considering how much people are willing to pay for the adjustment of certain characteristics. The stated method is most commonly used to determine value of the products whose energy characteristics have been improved, making them environmentally acceptable, such as cars and technology facilities that use alternative fuels. In this way it will be estimated whether consumers are willing to pay a specific (in most cases) higher price for the product of technology based on sustainable energy management.

Case study

Renewable electricity and sustainable development goals in the EU, World Development, Volume 125, 2020. https://www.sciencedirect.com/science/article/abs/pii/S0305750X19303419.

References

Afanasyeva, S., Saari, J., Kalkofen, M., Partanen, J., & Pyrhönen, O. (2016). Technical, economic and uncertainty modelling of a wind farm project. *Energy Conversion and Management, 107*, 22–33.

Ahmed, R., Sreeram, V., Mishra, Y., & Arif, M. D. (2020). A review and evaluation of the state-of-the-art in PV solar power forecasting: Techniques and optimization. *Renewable and Sustainable Energy Reviews, 124*, 109792.

Alla, S. A., Bianco, V., Tagliafico, L. A., & Scarpa, F. (2020). Life-cycle approach to the estimation of energy efficiency measures in the buildings sector. *Applied Energy, 264*, 114745.

Andersson, E., Arfwidsson, O., & Thollander, P. (2018). Benchmarking energy performance of industrial small and medium-sized enterprises using an energy efficiency index: Results based on an energy audit policy program. *Journal of Cleaner Production, 182*, 883–895.

Andrae, A. S. G., & Edler, T. (2018). On global electricity usage of communication technology: Trends to 2030. *Challenges, 6*(1), 117–157.

Araci, C., Haro, P., Giuntoli, J., & Ollero, P. (2017). Proving the climate benefit in the production of biofuels from municipal solid waste refuse in Europe. *Journal of Cleaner Production, 142*(4), 2887–2900.

Ballesteros, M., & Manzanares, P. (2019). Liquid biofuels. In *The role of bioenergy in the economy* (pp. 113–144). Academic Press.

Bedoić, R., Ćosić, B., & Duić, N. (2019). Technical potential and geographic distribution of agricultural residues, co-products and by-products in the European Union. *Science of the Total Environment, 686*, 568–579.

Berendes, D. M., Yang, P. J., Lai, A., Hu, D., & Brown, J. (2018). Estimation of global recoverable human and animal faecal biomass. *Nature Sustainability, 1*, 679–685.

Broin, E.Ó., Nässén, J., & Johnsson, F. (2015). Energy efficiency policies for space heating in EU countries: A panel data analysis for the period 1990–2010. *Applied Energy, 150*, 221–223.

Campton, M., & Rezaie, B. (2018). Investigating steam turbine feasibility to improve the sustainability of a biomass boiler using TRNSYS. *Sustainable Cities and Society, 43*, 86–94.

Castro Camioto, F., Moralles, H. F., Mariano, E. B., & Nascimento Rebelatto, D. A. (2016). Energy efficiency analysis of G7 and BRICS considering total-factor structure. *Journal of Cleaner Production, 122*, 67–77.

Chaudhary, P., & Rizwan, M. (2018). Voltage regulation mitigation techniques in distribution system with high PV penetration: A review. *Renewable and Sustainable Energy Reviews, 82*, 3279–3287.

Chebotareva, L., Strielkowski, W., & Streimikiene, D. (2020). Risk assessment in renewable energy projects: A case of Russia. *Journal of Cleaner Production, 269*, 122110.

Chen, F., Liu, L., Peng, J., Ge, Y., Wu, H., & Liu, W. (2019). Theoretical and experimental research on the thermal performance of ocean thermal energy conversion system using the Rankine cycle mode. *Energy, 183*, 497–503.

Colmenar-Santos, A., Rosales-Asensio, E., Borge-Diez, D., & Blanes-Peiró, J.-J. (2016). District heating and cogeneration in the EU-28: Current situation, potential and proposed energy strategy for its generalization. *Renewable and Sustainable Energy Reviews, 63*, 621–639.

Correa, D. F., Beyer, H. L., Fargione, J. E., Hill, J. D., Possingham, H. P., Thomas-Hall, S. R., et al. (2019). Towards the implementation of sustainable biofuel production systems. *Renewable and Sustainable Energy Reviews, 107*, 250–263.

Couto, T. B. A., & Olden, J. D. (2018). Global proliferation of small hydropower plants—Science and policy. *Frontiers in Ecology and the Environment, 16*(2), 91–100.

Crago, C. L., & Koegler, E. (2018). Drivers of growth in commercial-scale solar PV capacity. *Energy Policy, 120*, 481–491.
Danish, M., & Ahmad, T. (2018). A review on utilization of wood biomass as a sustainable precursor for activated carbon production and application. *Renewable and Sustainable Energy Reviews, 87*, 1–21.
Das, S. (2020). The National Policy of biofuels of India—A perspective. *Energy Policy, 143*, 111595.
Davood, F., Anda, M., & Shafiullah, G. M. (2020). Hydrogen production for energy: An overview. *International Journal of Hydrogen Energy, 45*, 3847–3869.
De Faria, F. A. M., Davis, A., Severnini, E., & Jarmillo, P. (2018). The local socio-economic impacts of large hydropower plant development in a developing country. *Energy Economics, 67*, 533–544.
Devlin, A. T., Pan, J., & Lin, H. (2018). Extended spectral analysis of tidal variability in the North Atlantic ocean. *JGR Oceans, 124*(1), 506–526.
Du, H., Matisoff, D. C., Wang, Y., & Liu, X. (2016). Understanding drivers of energy efficiency changes in China. *Applied Energy, 184*, 1196–1206.
Famoso, F., Prestipino, M., & Galvagno, B. A. (2020). Designing sustainable bioenergy from residual biomass: Site allocation criteria and energy/exergy performance indicators. *Applied Energy, 274*, 115315.
Feng, C., & Wang, M. (2018). Analysis of energy efficiency in China's transportation sector. *Renewable and Sustainable Energy Reviews, 94*, 565–575.
Filipović, S., Radovanović, M., & Golušin, V. (2018). Macroeconomic and political aspects of energy security. *Renewable and Sustainable Energy Reviews, 98*, 428–435.
Freire-González, J., & Vivanco, D. F. (2017). The influence of energy efficiency on other natural resources uses: An input-output perspective. *Journal of Cleaner Production, 162*, 336–345.
Fu, P., Qin, W., Bai, S., Yang, D., & Chen, L. (2019). Integrating large-area perovskite solar module with thermoelectric generator for enhanced and stable power output. *Nano Energy, 65*, 104009.
Gao, J., Guo, F., Ma, Z., Huang, X., & Li, X. (2020). Multi-criteria group decision-making framework for offshore wind farm site selection based on the intuitionistic linguistic aggregation operators. *Energy, 204*, 117899.
Gao, C., Na, H., Song, K., Dyer, N., Tian, F., Xu, Q., et al. (2019). Environmental impact analysis of power generation from biomass and wind farms in different locations. *Renewable and Sustainable Energy Reviews, 102*, 07–317.
García, R., Pizzaro, C., Lavín, A., & Bueno, L. J. (2017). Biomass sources for thermal conversion. Techno-economical overview. *Fuel, 195*, 182–189.
Geng, C., Yang, W., Sun, X., Wang, X., Bai, Z., & Zhang, X. (2019). Emission factors, ozone and secondary organic aerosol formation potential of volatile organic compounds emitted from industrial biomass boilers. *Journal of Environmental Sciences, 83*, 64–72.
Gollakota, A. R. L. (2018). A review on hydrothermal liquefaction of biomass. *Renewable and Sustainable Energy Reviews, 81*, 1378–1392.
González-García, S., & Bacceneti, J. (2019). Exploring the production of bio-energy from wood biomass. Italian case study. *Science of the Total Environment, 647*(10), 158–168.
Hamelin, L., Borzęcka, M., Kozak, M., & Pudetko, R. (2019). A spatial approach to bioeconomy: Quantifying the residual biomass potential in the EU-27. *Renewable and Sustainable Energy Reviews, 100*, 127–142.
He, P. Y., Zhang, Y. J., Chen, H., Han, C., & Liu, C. (2019). Low-energy synthesis of kaliophilite catalyst from circulating fluidized bed fly ash for biodiesel production. *Fuel, 257*, 116041.

Hicks, J., & Ison, N. (2018). An exploration of the boundaries of 'community' in community renewable energy projects: Navigating between motivations and context. *Energy Policy, 113*, 523–534.

Hrovatin, N., Dolšak, N., & Zorić, J. (2016). Factors impacting investments in energy efficiency and clean technologies: Empirical evidence from Slovenian manufacturing firms. *Journal of Cleaner Production, 127*, 475–486.

Hu, X., & Gholizadeh, M. (2020). Progress of the applications of bio-oil. *Renewable and Sustainable Energy Reviews, 134*, 110124.

Hunsberger, C., German, L., & Goetz, A. (2017). "Unbundling" the biofuel promise: Querying the ability of liquid biofuels to deliver on socio-economic policy expectations. *Energy Policy, 108*, 791–805.

Iddrisu, I., & Bhattacharyya, S. C. (2015). Sustainable Energy Development Index: A multi-dimensional indicator for measuring sustainable energy development. *Renewable and Sustainable Energy Reviews, 50*, 513–530.

Infelise, L., Kazimierczak, J., Wietecha, J., & Kopania, E. (2019). GINEXTRA®: A small-scale multipurpose modular and integrated biorefinery technology. In J. R. Bastidas-Oyanedel, & J. Schmidt (Eds.), *Biorefinery* (pp. 593–614). Cham: Springer.

Inshankov, O. V., Bogachkova, L. Y., & Popkova, E. G. (2018). Energy efficiency as a driver of global competitiveness, the priority of the state economic policy and the international collaboration of the russian federation. In O. Inshakov, A. Inshakova, & E. Popkova (Eds.), *Vol. 44. Energy sector: A systemic analysis of economy, foreign trade and legal regulations. Lecture notes in networks and systems* Springer.

Ioannou, A., Angus, A., & Brennan, F. (2017). Risk-based methods for sustainable energy system planning: A review. *Renewable and Sustainable Energy Reviews, 74*, 602–615.

Jackson, J. (2010). Promoting energy efficiency investments with risk management decision tools. *Energy Policy, 38*(8), 3865–3873.

Jakhongir, T., & Ugli, T. (2019). The importance of alternative solar energy sources and the advantages and disadvantages of using solar panels in this process. *International Journal of Engineering and Information Systems, 3*, 60–64.

Jamar, A., Majid, Z. A. A., Azmi, W. H., Norhafana, M., & Razak, A. A. (2016). A review of water heating system for solar energy applications. *International Communications in Heat and Mass Transfer, 76*, 178–187.

Jansen, M., Staffell, I., Kitzing, L., Quoilin, S., Wiggelinhuizen, E., Bulder, B., et al. (2020). Offshore wind competitiveness in mature markets without subsidy. *Nature Energy, 5*, 614–622.

Jordaan, S. M., Romo-Rabago, E., McLeary, R., Reidy, L., Nazari, J., & Herremans, I. M. (2019). The role of energy technology innovation in reducing greenhouse gas emissions: A case study of Canada. *Renewable and Sustainable Energy Reviews, 78*, 1397–1409.

Kabir, E., Kumar, P., Kumar, S., Adelodun, A. A., & Kim, K. H. (2019). Solar energy: Potential and future prospects. *Renewable and Sustainable Energy Reviews, 82*, 894–900.

Kabule, A., Ločmelis, K., & Blumberga, D. (2020). Analysis of the results of national energy audit program in Latvia. *Energy*, 117679.

Kannan, N., & Vakeesan, D. (2016). Solar energy for future world:—A review. *Renewable and Sustainable Energy Reviews, 62*, 1092–1105.

Khan, N., Kalair, A., Abas, N., & Haider, A. (2017). Review of ocean tidal, wave and thermal energy technologies. *Renewable and Sustainable Energy Reviews, 72*, 590–604.

Khosravi, A., Machado, L., & Nunes, R. O. (2018). Time-series prediction of wind speed using machine learning algorithms: A case study Osorio wind farm, Brazil. *Applied Energy, 224*, 550–566.

Klein, S. J. W., & Coffey, S. (2016). Building a sustainable energy future, one community at a time. *Renewable and Sustainable Energy Reviews, 60*, 867–880.

Kluczek, A., & Olszewski, P. (2017). Energy audits in industrial processes. *Journal of Cleaner Production, 142*, 3437–3453.

Kumar, Y. V. R., & Bhimasingu, R. (2017). Electrical machines-based DC/AC energy conversion schemes for the improvement of power quality and resiliency in renewable energy microgrids. *International Journal of Electrical Power & Energy Systems*, *90*, 10–26.

Lee, D., & Cheng, C. C. (2016). Energy savings by energy management systems: A review. *Renewable and Sustainable Energy Reviews*, *56*, 760–777.

Leijon, J., & Boström, C. (2018). Freshwater production from the motion of ocean waves—A review. *Desalination*, *435*, 161–171.

Liebe, U., Gewinner, J., & Diekmann, A. (2018). What is missing in research on non-monetary incentives in the household energy sector? *Energy Policy*, *123*, 180–183.

Liu, K. (2019). Effects of sample size, dry ashing temperature and duration on determination of ash content in algae and other biomass. *Algal Research*, *40*, 101486.

Locmelis, K., Blumberga, D., Blumberga, A., & Kubule, A. (2020). Benchmarking of industrial energy efficiency. Outcomes of an energy audit policy program. *Energies*, *13*(9), 2210.

Longati, A. A., Lino, A. R. A., Giordano, R. C., Furlan, F. F., & Cruz, A. J. G. (2018). Defining research & development process targets through retro-techno-economic analysis: The sugarcane biorefinery case. *Bioresource Technology*, *263*, 1–9.

Lotvi, S., Ma, W., Austin, K., & Kumar, A. (2019). A wet packed-bed scrubber for removing tar from biomass producer gas. *Fuel Processing Technology*, *193*, 197–203.

Lu, C. (2018). When will biofuels be economically feasible for commercial flights? Considering the difference between environmental benefits and fuel purchase costs. *Journal of Cleaner Production*, *181*, 365–373.

Lund, J. W., & Toth, A. N. (2020). Direct utilization of geothermal energy 2020 worldwide review. *Geothermics*, *90*, 101915.

Mami, M. A., Lajili, M., Khiari, B., & Jeguirim, M. (2020). Combustion of raw and densified Tunisian oleic by-products in a fixed bed reactor. *Fuel*, *277*, 118181.

Manzela, A. (2018). *General introduction to geothermal energy*. Geothermal Energy and Society, Springer.

Manzella, A., Bonciani, R., Allansdottir, A., Bolteghi, S., Donato, A., Giamberini, S., et al. (2018). Environmental and social aspects of geothermal energy in Italy. *Geothermics*, *72*, 232–248.

Maqbool, R. (2018). Efficiency and effectiveness of factors affecting renewable energy projects; an empirical perspective. *Energy*, *158*(1), 944–956.

Melikoglu, M. (2018). Current status and future of ocean energy sources: A global review. *Ocean Engineering*, *148*, 563–573.

Miglani, S., Orehounig, K., & Carmeliet, J. (2018). A methodology to calculate long-term shallow geothermal energy potential for an urban neighborhood. *Energy and Buildings*, *159*, 462–473.

Miranda, T., Montero, I., Sepúlveda Justo, D. J., Arranz, J. I., Rojas, C. V., & Nogates, S. (2015). A review of pellets from different sources. *Materials*, *8*(4), 1413–1427.

Mohanty, P., Muneer, T., Gago, E. J., & Kotak, Y. (2016). Solar radiation fundamentals and PV system components. In P. Mohanty, T. Muneer, & M. Kolhe (Eds.), *Solar photovoltaic system applications. Green energy and technology*. Cham: Springer.

Mossier, N. S., & Ileleji, K. E. (2020). Chapter 26—How fuel ethanol is made from corn. In *Bioenergy* (2nd ed.). Academic Press.

Motta, I. L., Miranda, N. T., Filho, R. M., & Maciel, M. R. F. (2018). Biomass gasification in fluidized beds: A review of biomass moisture content and operating pressure effects. *Renewable and Sustainable Energy Reviews*, *94*, 998–1023.

Moustakas, K., Loizdou, M., Rahan, M., & Nizami, A. S. (2020). A review of recent developments in renewable and sustainable energy systems: Key challenges and future perspective. *Renewable and Sustainable Energy Reviews*, *119*, 109418.

Munoz-Cupa, C., Hu, Y., Xu, C., & Bassi, A. (2021). An overview of microbial fuel cell usage in wastewater treatment, resource recovery and energy production. *Science of the Total Environment, 754*, 142429.
Neill, S. P., Hashemi, M. R., & Lewis, M. J. (2016). Tidal energy leasing and tidal phasing. *Renewable Energy, 85*, 580–587.
Nevzorova, T., & Kutcherov, V. (2019). Barriers to the wider implementation of biogas as a source of energy: A state-of-the-art review. *Energy Strategy Reviews, 26*, 100414.
Noorollahi, Y., Saeidi, R., Mohammadi, M., Amiri, A., & Hosseinzadeh, M. (2018). The effects of ground heat exchanger parameters change on geothermal heat pump performance—A review. *Applied Thermal Engineering, 129*, 1645–1658.
Ortega, M., del Rio, P., Ruiz, P., Nijs, W., & Politis, S. (2020). Analysing the influence of trade, technology learning and policy on the employment prospects of wind and solar energy deployment: The EU case. *Renewable and Sustainable Energy Reviews, 122*, 109657.
Özdemir, Ö., Hobbs, B. F., Van Hout, M., & Koutstaal, P. R. (2020). Capacity vs energy subsidies for promoting renewable investment: Benefits and costs for the EU power market. *Energy Policy, 137*, 111146.
Pan, S. Y., Gao, M., Shah, K. J., Zheng, J., Pei, S. L., & Chiang, P. C. (2019). Establishment of enhanced geothermal energy utilization plans: Barriers and strategies. *Renewable Energy, 132*, 19–32.
Pandey, W. C., Bajpai, O., & Singh, N. (2016). Energy crops in sustainable phytoremediation. *Renewable and Sustainable Energy Reviews, 54*, 58–73.
Pandey, A., Larroche, C., Gnansounou, E., Kumar, S., Claude-Gilles, K., & Ricke, D. S. (2019). *Biomass, biofuels, biochemicals* (2nd ed.). Elsevier.
Park, E., & Kwon, S. J. (2017). What motivations drive sustainable energy-saving behavior? An examination in South Korea. *Renewable and Sustainable Energy Reviews, 79*, 494–502.
Park, H., Lee, Y. J., & Park, J. (2018). Front and back TCO research review of a-Si/c-Si heterojunction with intrinsic thin layer (HIT) solar cell. *Transactions on Electrical and Electronic Materials, 19*, 165–172.
Paul, S., & Dutta, A. (2018). Challenges and opportunities of lignocellulosic biomass for anaerobic digestion. *Resources, Conservation and Recycling, 130*, 164–174.
Pelay, U., Luo, L., Fan, Y., Stitou, D., & Rood, M. (2017). Thermal energy storage systems for concentrated solar power plants. *Renewable and Sustainable Energy Reviews, 79*, 82–100.
Pérez-Lombard, L., Ortiz, J., & Velázquez, D. (2012). Revisiting energy efficiency fundamentals. *Energy Efficiency, 6*, 239–254.
Petrović, P., Filipović, S., & Radovanović, M. (2018). Underlying causal factors of the European Union energy intensity: Econometric evidence. *Renewable and Sustainable Energy Reviews, 89*, 216–227.
Proskurina, S., Jinginger, M., Heinimö, J., Tekinel, B., & Vakkilainen, E. (2019). Global biomass trade for energy—Part 2: Production and trade streams of wood pellets, liquid biofuels, charcoal, industrial roundwood and emerging energy biomass. *Biofuels, Bioproducts & Biorefining, 13*(2), 371–387.
Report and Data Portal. (2021). *National renewable energy national plans*. European Commission. https://ec.europa.eu/jrc/en/scientific-tool/nreap-data-portal.
Reuter, M., Patel, M. K., Eichhammer, W., Lapillonne, B., & Pollier, K. (2020). A comprehensive indicator set for measuring multiple benefits of energy efficiency. *Energy Policy, 139*, 111284.
Rimkus, A., Žaglinskis, J., Rapalis, P., & Skačkauskas, P. (2015). Research on the combustion, energy and emission parameters of diesel fuel and a biomass-to-liquid (BTL) fuel blend in a compression-ignition engine. *Energy Conversion and Management, 106*, 1109–1117.

Rosen, M. A., & Koohi-Fayegh, S. (2016). *Geothermal energy: Sustainable heating and cooling using the ground*. John Wiley & Sons, Ltd.
Roy, M. M., Wang, W., & Alawi, M. (2014). Performance and emissions of a diesel engine fueled by biodiesel–diesel, biodiesel–diesel-additive and kerosene–biodiesel blends. *Energy Conversion and Management*, *84*, 163–173.
Salpkari, J., & Lund, P. (2016). Optimal and rule-based control strategies for energy flexibility in buildings with PV. *Applied Energy*, *161*, 425–436.
Scarlat, N., Fahl, F., & Dallemand, J. F. (2019). Status and opportunities for energy recovery from municipal solid waste in Europe. *Waste and Biomass Valorization*, *10*, 2425–2444.
Sharaf, O. Z., & Orhan, M. F. (2014). An overview of fuel cell technology: Fundamentals and applications. *Renewable and Sustainable Energy Reviews*, *32*, 810–853.
Shi, R. S., Fan, X. C., & He, Y. (2017). Comprehensive evaluation index system for wind power utilization levels in wind farms in China. *Renewable and Sustainable Energy Reviews*, *69*, 461–471.
Shortall, R., Davidsdottir, B., & Axelsson, G. (2015). Geothermal energy for sustainable development: A review of sustainability impacts and assessment frameworks. *Renewable and Sustainable Energy Reviews*, *544*, 391–406.
Shurong, W., Gongxin, D., Haiping, Y., & Zhongyang, L. (2017). Lignocellulosic biomass pyrolysis mechanism: A state-of-the-art review. *Progress in Energy and Combustion Science*, *62*, 33–86.
Singh, D., Lakhi, K. S., Sil, S., Bhosale, S. V., Kim, I. Y., Albahily, K., et al. (2019). Biomass derived porous carbon for CO_2 capture. *Carbon*, *148*, 164–186.
Singh, R., Shukla, A., Tiwari, S., & Srivastava, M. (2014). A review on delignification of lignocellulosic biomass for enhancement of ethanol production potential. *Renewable and Sustainable Energy Reviews*, *32*, 713–728.
Singh, E., Tingas, E. A., Goussis, D., & Im, H. G. (2019). Chemical ignition characteristics of ethanol blending with primary reference fuels. *Energy & Fuels*, *33*(10), 10185–10196.
Soltani, M., Kashkooli, M., Souri, M., Rafei, B., Jabarifar, M., Gharali, K., et al. (2021). Environmental, economic, and social impacts of geothermal energy systems. *Renewable and Sustainable Energy Reviews*, *140*, 110750.
Stolarski, M. J., Szczukowski, S., Tworkowski, J., Krzyżaniak, M., Gulczyński, P., & Mleczek, M. (2013). Comparison of quality and production cost of briquettes made from agricultural and forest origin biomass. *Renewable Energy*, *57*, 20–26.
Sulaiman, C., Abdul-Rahim, A. S., & Ofozor, C. A. (2020). Does wood biomass energy use reduce CO2 emissions in European Union member countries? Evidence from 27 members. *Journal of Cleaner Production*, *253*, 119996.
Tallini, A., & Cedola, L. (2016). Evaluation methodology for energy efficiency measures in industry and service sector. *Energy Procedia*, *101*, 542–549.
Tammaro, M., Salluzzo, A., Rimauro, J., Schiavo, S., & Manzo, S. (2016). Experimental investigation to evaluate the potential environmental hazards of photovoltaic panels. *Journal of Hazardous Materials*, *306*, 395–405.
Taroun, A. (2014). Towards a better modelling and assessment of construction risk: Insights from a literature review. *International Journal of Project Management*, *32*(1), 101–115.
Thrän, D., Seidenberger, T., Zeddies, J., & Offermann, R. (2010). Global biomass potentials—Resources, drivers and scenario results. *Energy for Sustainable Development*, *14*(3), 200–205.
Trianni, A., Cagno, E., & Farné, E. (2016). Barriers, drivers and decision-making process for industrial energy efficiency: A broad study among manufacturing small and medium-sized enterprises. *Applied Energy*, *162*, 1537–1551.
Villena-Ruiz, L., Ramirez, J., Honrubia-Escribano, A., & Gómez-Lázaro, E. (2018). A techno-economic analysis of a real wind farm repowering experience: The Malpica case. *Energy Conversion and Management*, *172*, 182–199.

Waheed, R., Sarwar, S., & Wei, C. (2019). The survey of economic growth, energy consumption and carbon emission. *Energy Reports*, *5*, 1103–1115.

Wang, S., Jian, L., Shu, Z., Chen, S., & Chen, L. (2020). A high thermal conductivity cement for geothermal exploitation application. *Nature Resources Research*, *29*, 3675–3687.

Wang, M., Peng, J., Li, N., Yang, H., Wang, C., Li, X., et al. (2017). Comparison of energy performance between PV double skin facades and PV insulating glass units. *Applied Energy*, *194*, 148–160.

Wang, J., Zhang, L., Mao, Y., & Gong, F. (2020). An effective way to enhance energy output and combustion characteristics of Al/PTFE. *Combustion and Flame*, *2014*, 419–425.

Wu, Z., Feng, H., Chen, L., Tang, W., Shi, J., & Ge, Y. (2020). Constructable thermodynamic optimization for ocean thermal energy conversion system with dual-pressure organic Rankine cycle. *Energy Conversion and Management*, *210*, 112727.

Zabed, H., Sahu, J. N., Suely, A., Boyce, A. N., & Faruq, G. (2017). Bioethanol production from renewable sources: Current perspectives and technological progress. *Renewable and Sustainable Energy Reviews*, *71*, 475–501.

Zaporozhets, A. (2020). Methods and means for the control of the fuel combustion process. In *Vol. 287. Control of fuel combustion in boilers. studies in systems, decision and control*. Cham: Springer.

Zhang, J., & Xu, L. (2015). Embodied carbon budget accounting system for calculating carbon footprint of large hydropower project. *Journal of Cleaner Production*, *96*, 444–451.

Zhang, J., Xu, C., Song, Z., Huang, Y., & Wu, Y. (2019). Decision framework for ocean thermal energy plant site selection from a sustainability perspective: The case of China. *Journal of Cleaner Production*, *225*, 771–784.

Zheng, C. W., Wang, Q., & Li, C. Y. (2017). An overview of medium- to long-term predictions of global wave energy resources. *Renewable and Sustainable Energy Reviews*, *79*, 1492–1502.

CHAPTER 8

Energy security

Contents

The concept of energy security

Energy security is one of the development priorities of each country. Therefore, each national economy strives to provide a satisfactory level of energy security and to conceptualize, plan, and implement it following its own needs and specifics (Filipović, Radovanović, & Golušin, 2018).

Initial reflections on energy security arose in the 1970s as a result of oil crises. In October 1973, the first oil crisis began with members of the Organization of Arab Petroleum Exporting Countries (OAPEC) imposing an embargo on oil exports to the Netherlands, the United Kingdom, Portugal, Japan, Canada, the United States, Rhodesia, and South Africa. The reason for the embargo was the Yom Kippur War, which was not energy-related. Nevertheless, this event marked a new era in which oil started being used as an oil weapon, i.e., this event started the practice of using oil as leverage for political gains (Akins, 1973). Consequently, the price of a barrel of oil rose from USD 3 to 12 globally. As the sudden rise in the price of oil endangered the security of supply as well as the economic development of the countries, in 1974, the International Energy Agency (IEA) was formed with the initial goal of responding to physical oil supply disruptions. Over time, the IEA has become an international source for energy statistics, an advisor on energy policymaking along with issues of energy security (Paravantis, Kontoulis, Ballis, Tsirigotis, & Dourmas, 2018).

Sustainable Energy Management
https://doi.org/10.1016/B978-0-12-821086-4.00005-2

The problem of energy security became commonplace, as the second oil shock in 1979 followed very quickly, which was politically motivated by the escalation of the Iranian Revolution, and resulted in a double rise in prices over 1 year.

Although energy security issues were initially raised due to the oil crises, energy security was later linked to the Gulf War as well as to other crises (terrorist attack on September 11, 2001, in the United States), natural disasters (Hurricane Katrina in 2005), and conflict situations (the Russia-Ukraine gas dispute in 2005–06). Today, energy security is a matter of national security, and due to the different availability of energy resources, different political systems, economic welfare, ideologies, geographical locations, and international relations, energy security can mean different things and the priorities may vary.

Thus, the term energy security is of a relatively recent date and has undergone an upgrade over the last few decades. Initially, the term energy security meant, above all, technical security of supply, i.e., the provision of all necessary technical preconditions for the supply of sufficient quantities of energy (Sovacool & Mukherjee, 2011). During the 1970s and early 1980s, during several oil crises, oil price volatility directly affected energy security, so in addition to the technical dimension, the economic prerequisites for providing imported energy were pointed out. In the early 1990s, after the war in the Persian Gulf region, energy became an unavoidable subject of geopolitical changes. Hence, energy security gained a geopolitical component. The transition to the 21st century brought the need for environmental protection so that the concept of energy security has gained its ecological dimension. At the same time, there is a growing belief that access to energy is the right of every human being and one of the essential preconditions for an adequate quality of life, so energy security includes elements of the social dimension as well as.

In defining energy security, the IEA has developed a short-term and long-term approach to energy security. It views the short-term approach of energy security exclusively as the ability of a country's energy system to respond immediately and in the best possible way to changes in the balance between supply and demand for energy. On the other hand, the long-term approach to energy security focuses on finding and implementing supplies that are in line with economic development, with the need to preserve the quality of the environment. The IEA defines energy security as the uninterrupted availability of energy sources for all needs, at an affordable price (Ang, Chong, & Ng, 2015).

The Asia Pacific Energy Research Center (APERC) formulates the same ideas with further focus stating that energy security is the ability of an economy to guarantee the availability of energy resource supply in a sustainable and timely manner with the energy price being at a level that will not adversely affect the economic performance of the economy. Whereas environmental sustainability was not directly incorporated in IEA's definition except for its related long-term dimension, the APERC explicitly mentions it (Report: A quest for energy security in the 21st century. Asia Pacific Energy Researchz Centre, Tokyo, Japan, 2007).

The European Commission agrees as it defines energy security as the uninterrupted physical availability of energy products on the market, at a price that is affordable for all consumers (private and industrial), while respecting environmental concerns and looking toward sustainable development (Radovanović, Filipović, & Golušin, 2018).

On the other hand, the United Nations Development Program (UNDP) defines energy security as "part of national security, which implies a continuous supply of energy in various forms, in sufficient quantities, and at an affordable price" (Report: World energy assessment: Energy and the challenge of sustainability, UNDP, 2015).

The IEA, APERC, UNDP, and European Commission's views on energy security are thus focused on the following three elements:

1. The physical availability and accessibility of supply sources,
2. The economic affordability and
3. The long-term environmental sustainability.

This approach is complex because it encompasses three basic dimensions: availability, reliability, and acceptability. The availability of energy goods and services implies the physical disposal of the required quantities of energy, by type (with adequate stocks), assuming the functioning of an adequate world energy market (Cherp & Jewell, 2014). Reliability shows how resistant the energy system is to disruption because the uninterrupted flow of energy can be interrupted due to technical problems and accidents, natural disasters, pandemics, changes in the world energy market, changes in the geopolitical sphere, actions of hostile groups or individuals, high-tech crime and other challenges, risks, and threats. Acceptability depends on the complex relationships that exist in the world energy market and is reflected in national energy security, even in countries that have large natural energy reserves.

There is a clear division on the acceptability of short-term and long-term definitions of energy security. Proponents of the short-term approach believe that energy security is a matter of the ability of the energy sector (and all relevant systems) to provide enough energy for the needs of one country (Augustis, Krikštolaitis, Martišauskas, Pečiulytė, & Žutautaitė, 2017). Security of supply is in the focus of observation. Although there is clear evidence that energy security is not just security of supply, most modern measurement methods are based precisely on determining the degree of security of supply, without considering environmental, social, and other aspects. Of course, the security of supply is highly important. However, it is only one aspect of energy security, which in no way measures the related environmental parameters and the welfare of citizens (Narula, Reddy, Pachauri, & Dev, 2017).

In addition to these four institutions, some other authors have contributed to the most comprehensive definition of energy security. Yergin (1988) for the first time expanded the concept by introducing a geopolitical dimension, that is, defining energy security as "assurance of an adequate, reliable supply of energy at a reasonable price and in ways that do not jeopardize major national values and objectives."

Based on this, we can say that the broader concept of energy security implies the country's ability to provide enough energy (by type, quantity, and quality), in the necessary place, at the required time, which will be sufficient to meet the needs of the economy and society, without endangering the environment, at an acceptable price, and with the provision of adequate reserves.

Energy security and energy risk

Although there is no single accepted definition of energy security, it is evident that energy security depends on a large number of environmental factors that affect the energy supply chain. Due to the difficulty in measuring all these risks simultaneously, the concept of energy security is mainly reduced to the analysis of one or several dimensions in the literature.

Risk as the probability that a particular event will occur is a complex phenomenon and can be studied and observed from several points of view. Thus, depending on the criteria, we classify risks according to the source of risk, the extent of impact, and various aspects of risk such as speed, size, endurance, spread, singularity, or certainty of impact (Akhtaruzzaman, Boubaker, Chiah, & Zhong, 2021).

If the source of risk is observed, then a rough classification into three risk categories can be made: technical risks, human risks, and natural risk sources.

Technical risk refers to various infrastructure problems and constraints such as dilapidated and outdated infrastructure capacities that in themselves jeopardize the process of energy production, transmission, and distribution, and can also result in high losses in the transmission and distribution network (Boyko, Rudnichenko, Kramskoy, Hrechukha, & Shibaeva, 2017). Besides, technical risk may relate to inadequate network of oil and gas pipelines, inadequate protection of plants, insufficiently developed infrastructure, etc. Human risk can refer to various types of human error and negligence, but also high fluctuations in energy demand, underinvestment in energy capacities, various forms of sabotage and terrorism, political and geopolitical instability, wars, embargoes on exports and imports of energy. Natural risk sources refer to the problem of depletion of fossil fuels, the problem of insufficient availability of renewable energy sources, the occurrence of natural disasters, etc. Natural and human risk sources are often viewed in the context of security of supply and are associated with the "availability" and "accessibility" of energy resources (Aven, 2016). Sources of risk and scope of impact measure are presented at Fig. 8.1 (Winzer, 2012).

If the scope of the impact measure is observed, i.e., how energy security is viewed, one of the important aspects is to consider the risk of sudden changes in the availability of supply concerning demand. Continuity in energy supply can refer to different phases of the process of transformation of primary energy, i.e., supply to the final consumer, and it can depend on the price and availability of oil, gas, coal, or electricity. Another risk may relate to continuity in supply services, i.e., prices, and availability of heating, transmission, distribution services, etc. (Thorbecke, 2019). Also, if the scope of the impact measure is observed, human safety and environmental sustainability should be taken into account.

Many authors further restrict the concept of energy security by introducing subjective severity filters to distinguish between secure and insecure levels of continuity. The severity of a threat increases with the speed, size, sustention, and spread as well as with decreasing singularity and sureness of the impacts. Speed refers to the time frame during which risk can materialize. Accordingly, there is a constant scarcity—(e.g., potency in renewable energy sources), and fast shocks: supply disruption due to war, technical disasters, or natural disasters (Murshed & Tanha, 2021).

The size of threat impacts refers to the extent of changes in scarcity within the affected area. In that sense, there are impeding changes (indicating the probability of a negative impact without a direct impact on consumers), small changes (affecting consumers, but not the system), and

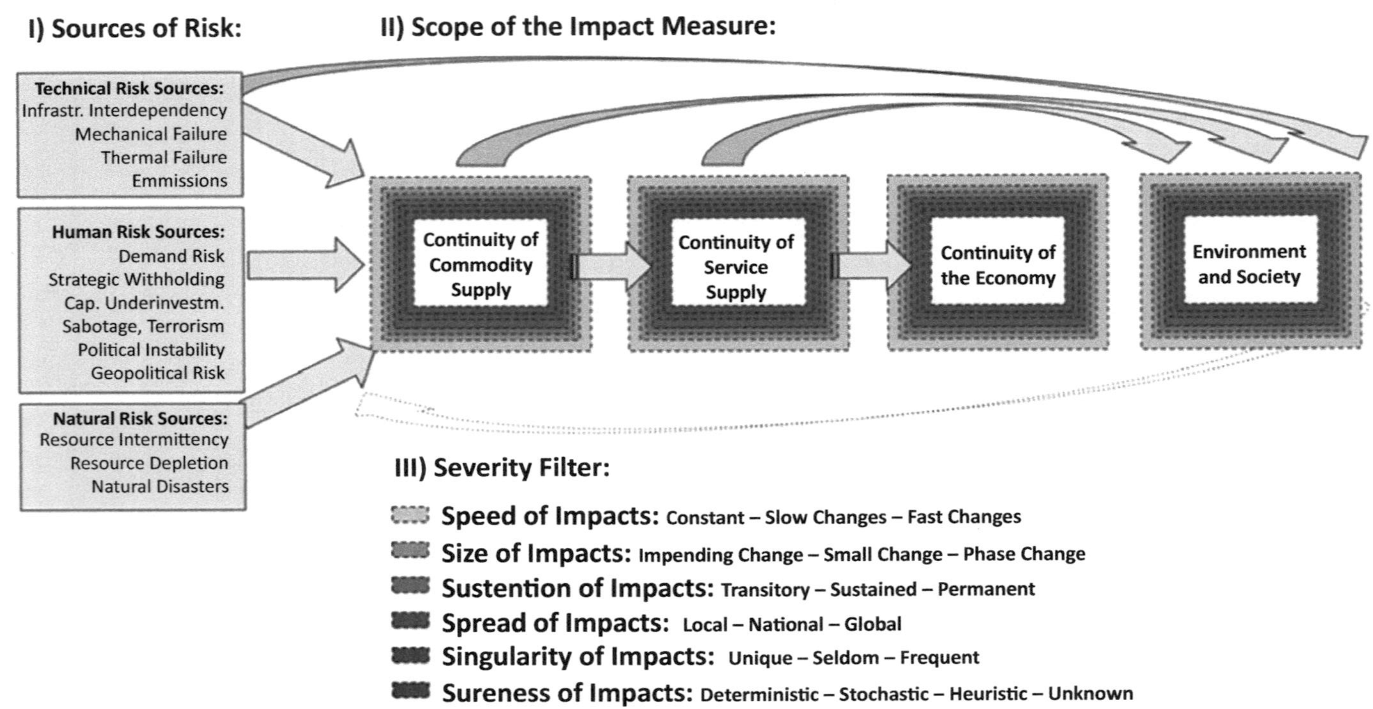

Fig. 8.1 Dimensions of energy security.

phase changes (affecting consumers, but also the system). A good example of phase changes could be physical barriers to energy supply or global warming (Kester, 2021).

Depending on the spread of threat impacts, risks differ at the local, national, and global levels. Endangered energy security due to technical problems is an example of local risk. At the national level, an example of endangered energy security is political risk or an embargo in energy trade. Since global warming occurs at the supranational level and has effects on the whole world, the risk of global warming is an example of the risk of the highest rank.

The singularity of threats refers to the frequency of occurrence of the risk. In that sense, we make a distinction between unique (e.g., nuclear catastrophe), infrequent (political conflicts, natural disasters), and frequent (technical problems) threats. As a result, the literature states operational and strategic risks, where the former include frequent threats, whereas strategic risk refers to infrequent/rare treats (Wang & Zhou, 2017).

Risks are also classified according to the level of predictability (Hammoudeh, Makni, Ben-Salha, & Ajmi, 2021). Accordingly, there are four types of risk: predictable (example of energy resource depletion), probabilistic (example of a technical constraint), heuristic (possibility of political conflict or terrorist attack), and completely unpredictable (anthropogenic global warming).

Based on all mentioned, it can conclude that energy management that implies achieving the desired level of energy security is a complicated and time-consuming process. This process is under the influence of a lot of different factors, and at its center is the energy sector (exploitation, transport, and energy consumption). The energy sector operates in an environment that is more or less unstable, whereby the environment that affects the processes of obtaining, distributing, and consuming energy in the world is especially variable and difficult to predict (Matsumoto & Shiraki, 2018). Namely, there are numerous possibilities for predicting the movements on the global energy market, but the specifics and significance that energy has for the whole of humanity necessarily lead to problems and events that are difficult to manage. Due to all the above, energy management is a process that is associated with increased risk and therefore must be planned and implemented with special attention (Tian et al., 2018).

Sustainable energy management is, in essence, an additional complication of the energy management process and therefore necessarily becomes an even more complex process when it comes to risk assessment that accompanies and affects it. In the process of sustainable energy management, risk is a separate process, but one that nevertheless takes place integrated with the process of sustainable energy management, intending to achieve and maintain the desired level of energy security (Malkova & Malkov, 2018).

When assessing the risk in the process of sustainable energy management, it is necessary to take into account and analyze a large number of internal and external factors. Also, it is essential to analyze economic and noneconomic risk factors that may affect the success of the implementation of sustainable energy management in a particular business system.

Internal risk factors include a large group of events that may occur in the economic environment, which primarily include (Spada & Burgherr, 2020):

- procurement market—where it is necessary to assess all factors of energy supply, their availability, purchase price, distribution conditions, quality, specifics, with a particular analysis of the possibility of supplying energy produced from alternative sources. Also, it is necessary to analyze the options for energy production from own sources;
- competition—every company operates in an environment where there is competition. It is necessary to consider the strength and quality of competition, its number, geographical distribution, and market segments it covers. Modern companies are fighting for consumers, among other things, by influencing the increase in the level of energy security of the country, so that the acquisition of a strong competitive position largely depends on the behavior toward the use of energy resources;
- legal regulations—in each country certain legal regulations determine the rules of business so that any change can lead to significant problems for the security of the state. Modern legislation mostly supports sustainable energy management, but the risks of change in this area are still possible; and
- technical and technological innovations—are of great importance in the process of security energy management. All changes in the technology market have to be specifically anticipated. Also, it is necessary to take into account the quality, prices, availability, specifics of delivery, and installation of a particular technology, as well as all other factors of this type.

External risk factors include a range of diverse factors that can positively or negatively affect the energy sector to a large extent. These factors need to be analyzed as accurately as possible due to the specificity of the global distribution and the importance of energy in the development of all humankind (Franc-Dąbrowska, Mądra-Sawicka, & Milewska, 2021):

- availability of energy resources—uneven distribution of energy resources put a large number of economies in the world in an unequal position because many countries have to import energy. Therefore, each country strives to provide sufficient amounts of energy for its functioning but is highly dependent on the rules dictated by countries that have energy resources.
- conditions of natural disasters—energy resources are products of a country, which have a high possibility of being exposed to natural disasters, which can jeopardize the ability to supply the world market to a great extent.
- the conditions for the occurrence of disasters due to human activities—energy production and processing plants, as well as facilities and means of distribution, are inherently complex and as such may be particularly susceptible to industrial disasters of smaller or larger scale.
- political factors—are factors that largely determine the functioning of the world today, especially in the sphere of energy distribution, which is a limited resource. We must consider all political changes with special attention to political predictions to minimize the risk of their occurrence.
- state of war and other armed conflicts—the risk of armed action is always present in regions with large energy reserves, which poses a constant risk to the availability of energy for all those who depend on energy imports.

Only the basic factors that need to be taken into account when assessing the risk of operation of power plants are listed. Their analysis is complex but necessary because the transition from the traditional to the safe way of energy management is a kind of strategic change that necessarily leads to large and long-term changes. Risk assessment involves performing a series of evaluations that are exceptionally complex when it comes to activities related to energy production and consumption because there is a high possibility of environmental incidents with major consequences (Sýkora, Marková, & Diamantidis, 2018).

In addition to the harmful consequences for the environment, the negative consequences for the energy sector and the country as a whole should certainly be considered, because all adverse events in the energy sector can have an impact on the broader social, economic, or political environment. Plants that deal with the production and transportation of energy conventionally are plants that carry a high-security risk. A particular problem for risk assessment is producing energy in nuclear plants.

Risk analysis, as part of the risk management process, involves the analysis of each critical point identified. During this phase, it is necessary to determine all the parameters of the observed risk (Azzuni & Breyer, 2018):

- the origin of the risk, where it is necessary to determine why a critical point occurs at a certain place in the technological process, that is why pollution occurs. Proper assessment of the origin of pollution is very important because it can provide valuable information that can immediately offer adequate solutions. The situation is different if it turns out that the cause of harmful effects is a human factor, malfunctions in the installation of the plant, weak technological solution, and the like;
- type of pollution, implies the detection of the nature of pollution, that is, its physical, chemical, or biological nature;
- direction of action, represents different pollutants directed to diverse natural systems, some pollute the air, some water, some directly endanger human health, and most often act in a complex manner;
- intensity of action implies defining the strength of the impact of pollutant emissions because it certainly does matter whether it is a small or mass contamination;
- hazard as a final analysis that indicates the real danger of the polluting factor to the environment.

After assessing the characteristics of the risk event itself, a list of priorities is determined. Energy consumption is often associated with risks of varying intensity but is essentially the cause of a large number of risks that exist in performing a particular activity. Energy consumption is a process that necessarily causes pollution and therefore requires special attention (Papież, Śmiech, & Frodyma, 2019).

Risk assessment, where all previously determined parameters are considered together and a judgment is made as to whether an individual critical point of high, medium, or low risk is present. Further procedures depend on this assessment. High-risk impacts should be eliminated as soon as possible, while low-risk impacts can in most cases be eliminated more simply. Experience shows that goals must be ranked according to the degree of urgency and ability to be solved and that they must then be approached individually. Only when one security problem that arises in the energy sector is solved, another problem can be solved (Wieczorek-Kosmala, 2020).

Risk management involves a series of activities (implementation of protection measures, employee training, strict and constant control) that are carried out to properly deal with critical points that have been detected in previous procedures. Some technologies and procedures are associated with risks, but with proper management, the possibility of incidents is minimized.

The risk assessment may establish that a particular technology has a large number of critical points and surpassing them is extremely expensive and therefore unprofitable. In such a case, it is indicative to think about changes in the technological process or about a change of the entire technology. Modern technologies are, as a rule, energy-efficient and safe, and energy consumption is controlled. However, in many parts of the world, modern technologies are not available, and the problem of transport, which is necessary for performing most human activities, still exists as an activity that consumes fossil fuels and replacements are not yet sufficient or adequate (Wang, Mao, Chen, & Yu, 2018).

Today, the assessment of the risk of performing a certain activity is a necessary parameter when evaluating future investment activities, that is, making decisions on starting a new activity. Every business plan or investment project in the energy sector must provide information on possible environmental risks, as well as offer information on how the company intends to keep risks under control.

Risk is an event that can be different and intense. Risk assessment is essentially a complex and multifactorial analysis that is related to all aspects of business and is especially analyzed to assess the economic feasibility of business and security of energy infrastructure as a whole. Namely, certain risks can be so high as to jeopardize the survival of the company itself or require investments that are so large that they impose the need for a strategic turn in business, which is an important decision with far-reaching consequences. Modern business rules require an assessment of work risk, especially when it comes to energy consumption, both in existing companies and when investing in new activities. There are three fundamental issues to consider before considering ranking energy projects (Iannou, Angus, & Brennan, 2017).

- It is first necessary to assess the strategic scope of the analyzed project or endeavor, that is, to evaluate how strategic and how opportunistic the project should be. Surely, projects with a significant strategic character should be the subject of a more detailed analysis. Most projects related to energy consumption can be considered as projects of this type because energy is a strategic resource and all activities in this field are of importance.
- Secondly, it is necessary to analyze the type of risks that exist. In conventional management, it is understandable that all investments,

regardless of whether they are potentially profitable, can carry a significant risk of failure. That is especially true in the case of investments in the energy sector whose potential profitability may depend on uncontrolled risks that are not evenly distributed in the observed area.

- Thirdly, the economic implications of the project need to be taken into account. Regardless of the priority given to sustainable energy management, ways to achieve equality or equity in balancing resources on condition to achieve maximum profitability per unit of money spent should be explored. It is necessary to assess whether the indicators should be adjusted to justify forced projects to achieve goals that are not narrowly economically oriented and safe. Energy production and energy resource management is an issue of strategic importance for each country, so these types of projects cannot be viewed solely as economic ventures.
- Finally, it is especially important to take into account the sensitivity of the energy sector to cyber-attacks.

Assessing the potential risk posed by the implementation of a particular project is an exceptionally complex undertaking, and evaluating energy risk makes it even more complicated. The issue is complicated further by global problems that deepen the risk in this sector. Namely, the problems of uneven distribution of energy resources, conditioning of energy supply, uncertain market, and price fluctuations are just one of the risk factors that should be taken into account. An overview of the factors to consider are the following:

- Exposure—it is necessary to assess how much and which type of uncontrolled risk factors can affect potential results.
- Timelines—it is necessary to consider whether there is a trend for important risk factors to become more or less favorable. If the time factor is misjudged, the project may fail due to premature investment or become profitable over too long a period. Most projects in the energy sector are expensive, and the demands in this direction are expected to be even stronger.
- Volatility—it is necessary to assess the probability of the expected result if all uncontrolled risk factors act. In the case of analysis of this factor, it is essential to have an adequate risk management plan because this factor has an extremely devastating effect if all types of risk occur at the same or in a short time. When it comes to energy projects, a relatively high level of these impacts can be expected, which is a consequence of the instability in the global energy market; therefore, it is advisable to perform this type of analysis with special attention.

- Safety—regardless of the number of risks, it is necessary to consider whether there is a way to minimize or mitigate the impact of uncontrolled risk factors. That most often refers to a step in the risk management process when making a decision.
- Persistence—projects in the energy sector are usually long-term, so it is necessary to assess how long uncontrolled risk factors will affect the success of the project. In this way, the possibility of terminating the observed risks during the project can be considered. Also, it is necessary to predict whether the effect of uncontrolled risk factors will continue after the end of the project life.

Only the fundamental factors that should be taken into account are listed, e.g., when planning investments and operations in the energy sector. As energy investments represent capital investments, risk analysis is a key phase of the strategic decision-making process with extraordinary emphasis on risk management in the field of protection against cyber attacks on the energy sector (Report: Cyber security in the energy sector. Recommendations for the European Commission on a European strategic framework and potential Future legislative acts for the energy sector, 2017).

Energy security measurement methodologies

Following different energy security definitions and respecting risks, decision-makers need to be able to derive effective policies for their countries. Hence, as a first step, they need to be able to quantify the energy security of their country to situate their performance and compare it with the one from other countries. Given the multiple dimensions of energy security, to quantify energy security, indicator-based approaches that include several indicators at the same time are most commonly used to provide a multidisciplinary approach. The Multicriteria Decision Analysis method is used, based on which several individual indicators are combined into an overall score, usually called an index or composite indicator, which exhibits the performance of the alternatives. Thanks to the effective communication tool for policymaking that indices offer, their popularity has been increasing recently. Numerous methods to combine individual indicators into an index are reported in the literature; as a consequence, several different indices have been developed (Gasseer, 2020).

After 2000, several approaches for measuring energy security were developed to quantify energy security, which would enable more efficient monitoring, comparison, and prediction. Currently, the most commonly used methods for quantifying energy security are:

- Shannon-Wiener Index,
- Energy Security Index,
- Supply/Demand Index for long-term security of supply,
- Oil Vulnerability Index,
- Vulnerability Index,
- Risky External Energy supply,
- Socioeconomic Energy risk,
- The US/International Energy Security Risk Index,
- MOSES—The IEA model of short-term energy security,
- AESPI—Aggregated energy security performance indicator.

Shannon-Wiener Index is a relatively simple way to determine the degree of energy security. The index determines the degree of diversification of the energy market, based on the share of each type of energy and the part of the market occupied by a particular supplier.

$$H' = -\sum_{i=1}^{R} p_i \log p_i \tag{8.1}$$

where, p_i is the share of a particular energy source; i is the energy mix or share with a particular supplier i.

A higher value of the index indicates greater diversification of the energy market, which presumably provides the basis for a higher degree of energy security. This index is an extremely valuable indicator of the degree of market diversification and, as such, mostly indicates the (in) dependence of a particular country concerning energy suppliers. Higher market diversification indicates the ability of a country's energy supply not to be strongly influenced by external changes and to adapt relatively smoothly in the event of a change. The main objection to this method of calculation is that the said index is solely oriented to the supply of energy and as such, does not take into account the remaining environmental parameters.

Energy Security Index is focused on measuring the causes of energy instability and uncertainty by observing the prices and physical availability of energy:

$$ESI = \sum_i \left(r_i \cdot S_{i,\text{oil}}^2\right) \cdot c_{\text{oil}}/TPES + \sum_i \left(\text{r}_\text{i} \cdot S_{i,\text{gas}}^2\right) \cdot c_{\text{gas}}/\text{TPES}$$

where, $s_{i,\text{oil}}$ is share of each supplier (region) i in the oil market, r_i is ranking the region by degree of political risk i, and $C_{\text{oil}}/TPES$ is share of total primary energy supply.

The above index shows the share of oil and gas in the total energy supply, the share of a particular supplier in the market, and the political risk that exists in the supplier country. Like the previous ones, this index is oriented exclusively on security of supply. Accurate risk measurement in the supplier country is, in turn, also questionable.

Supply/Demand Index for long-term security of supply—S/D Index observes the degree of sensitivity and flexibility both on the supplier side and on the demand side. S/D is a composite indicator composed of over 30 individual indicators. Indicators that look at the demand side show the characteristics of consumption. Indicators looking at the supply side show the features of demand in all sectors. Fig. 8.2 shows the elementary indicators that make up this index.

The S/D Index is composed of a relatively large number of indicators, which is a prerequisite for a high degree of measurement accuracy. However, its application is limited to a small number of countries, where there is a quality data collection system.

Oil Vulnerability Index is a type of composite indicator, which observes the following parameters: the ratio of oil imports to GDP (gross domestic product), oil consumption per unit of GDP, GDP per capita, share of oil in total energy supply, share of domestic oil reserves in total oil consumption, dependence from oil imports, diversification of supply sources, political

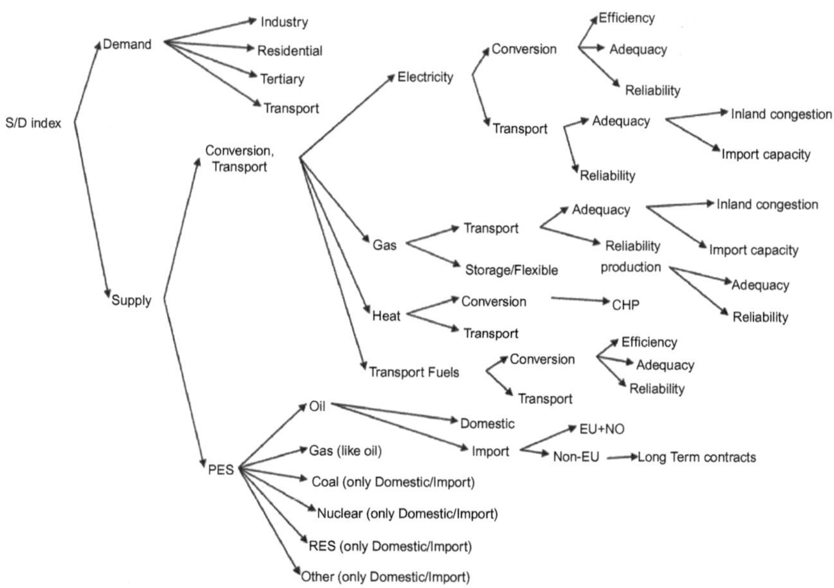

Fig. 8.2 Supply/Demand Index for long-term security of supply.

risks in supplier countries, and market liquidity. This index, like the previous one, is oriented exclusively on security of supply, looks only at the degree of vulnerability of the country to problems in oil supply, and does not look at other energy sources.

Vulnerability Index is a composite index that includes five indicators: energy intensity of GDP, energy import dependence, ratio of energy-related carbon emissions to TPES, electricity supply vulnerability, and nondiversity in transport fuels. The specified index introduces an environmental parameter (includes carbon emission). It considers only sensitivity in the field of electricity supply and diversification of transport fuels as proxy measures.

Risky external energy supply is based on the security of supply and assesses the risk of imports, i.e., the assessment of import dependence through the determination of the degree of diversification. It pays great attention to transport safety and assumes that transporting energy from a greater distance is riskier for the importing country.

$$REES_a^f = \left[\Sigma_i \left(\frac{NPI_{ai}^f}{NPI_a^f} \right) F_{ia}^f r_i d_{ia} \right] \cdot NID_a^f \cdot SF_a^f$$

where, NPI_{ai}^f is a positive energy sources import f from the country i to the country a; NPI_a^f is the sum of imports from all suppliers a; F_{ia}^f is the substitutability of energy imports f from the country i to the country a; r_i is political risk index in the energy-exporting country; d_{ia} is a measure of distance between countries i and a; NID_a^f is import dependence of the country a for energy f; and SF_a^f is share of energy f in the country a.

Socioeconomic Energy Risk Index is focused mainly on the assessment of total risk and includes the following indicators: diversification of energy sources, diversification of suppliers, energy resources, energy transport, energy intensity, dependence on energy imports, substitutability of energy sources, political stability, market liquidity, and GDP. Like any other index that looks at a large number of indicators, it can be used with great accuracy only in developed countries, where there is a reliable data collection system. Also, it uses the political stability indicator, the determination of which is associated with a large number of obstacles for a more accurate measurement of energy security.

The International Energy Security Risk Index was created in 2012 on the basis of the US index as an attempt to compare the level of US energy security with the level of energy security with other developed countries (25 countries in total). The mentioned index pays special attention to the characteristics of the world energy market. The number of individual indicators has been reduced from 83 to 29, with eight specific areas being observed. They are (Index of U.S. Energy Security Risk, 2018):

- Global indicators for crude oil: the security of world oil reserves, the security of world oil production, the security of world natural gas reserves, security of world natural gas production, the security of world coal reserves, security of world coal production.
- Indicators of energy imports: imports of crude oil, imports of natural gas, imports of coal, total imports of energy, costs of energy imports per unit of GDP.
- Energy import costs: intensity of energy import costs, energy import costs per capita, retail electricity prices, and crude oil prices.
- Price and market instability: volatility of crude oil prices, the variability of energy import costs, and use of world refineries.
- Energy use: energy consumption per capita, energy intensity, and oil intensity.
- Electricity sector: diversification of capacities for electricity production and production of energy from renewable sources.
- Transport sector: energy consumption per capita for transport and transport intensity.
- Environmental indicators: trend of carbon dioxide emissions, carbon dioxide emissions resulting from energy consumption per capita.

MOSES model observes only short-term energy security, expressed in days or weeks. Energy security is a measure of three basic indicators. They are good conditions (adequacy and reliability of energy resources and energy infrastructure), sovereignty (sensitivity to external threats), and resilience (ability to withstand and respond to various disturbances). The model uses 35 different indicators, but does not take into account indicators that have long-term consequences (environmental and social) and excludes estimates of availability and changes in energy prices (Podbregar, Šimić, Radovanović, Filipović, & Šprajc, 2020).

Energy Trilemma Index

Each of the approaches for defining and measuring energy security has its specifics, advantages, and disadvantages. Of all the indices, only a few have a comprehensive scope, and they attempt to measure all the factors of energy security. Among those developed by institutions and most frequently referred to by the professional public are the energy Trilemma Index developed by the World Economic Forum and the International Index of Energy Security risk developed by the Institute for 21st century energy.

Given the current trends, it is necessary to adopt the view that measurement methods must be continually re-examined and modified in line with the dynamic changes in the world geopolitical scene. Thus, for example, the emergence of the global economic crisis and the sky-high rise in oil prices just before the outbreak of the global economic crisis harmed the energy security of importing countries, and the growth of the foreign trade deficit affected macroeconomic stability. On the other hand, the dramatically low price of oil at the beginning of 2016 brought with it not only some disturbances on the energy market, which affected the macroeconomic stability of countries, but geopolitical consequences as well.

An aggravating circumstance is the fact that many of the factors relevant to energy security are not quantitatively measurable. Thus, for example, geopolitical circumstances, political instability, and war conflicts affect the price of energy and hence energy security. However, it is hard to quantify political instability in countries. Also, it should be taken into account that energy security is a dynamic category that monitors global change. Therefore, it is necessary to monitor not only global changes but also changes in the value of the defined energy security indicator.

With all this in mind, a proposal for a new Energy Security Index (ESI) was presented, which would include factors that reflect political, economic, and environmental aspects. It is a composite index that was tested on a sample of EU-28 member states for the period from 2003 to 2014.

Guided by the EU priorities in the field of energy policy, energy intensity, reduction of CO_2 emissions, and share of renewables in energy consumption were taken into account when defining the ESI. Besides, the index includes other indicators for energy security such as energy dependence, GDP per capita, final energy consumption per capita, electricity consumption per capita, and Euromoney country risk (ECR). ECR is an indicator that includes political and economic aspects, and its introduction into the ESI index is significant because it looks at the sociopolitical aspect.

Namely, energy security should include geopolitical turmoils and dynamic changes in the global economic environment. Also, it should take into account the fact that political and social aspects are not quantitatively measurable and that there is a general problem with high-quality data on political dimensions. The solution was in the ECR index, which is calculated based on the application of qualitative and quantitative methods. The qualitative part of the index (weight 70%) was obtained based on a survey of expert opinions around the world regarding political risk, economic performance, and structural assessment. Those three aspects are covered by 15 indicators. The categories of political risk scored are as follows: corruption; government nonpayments/nonrepatriation; government stability; information access/transparency; institutional risk; regulatory and policy environment. Economic risk is covered by parameters such as bank stability/risk; GNP outlook; unemployment rate; government finances; monetary policy/currency stability. The categories of structural risk scored are as follows: demographics; hard infrastructure; labor market/industrial relations; soft infrastructure. Quantitative indicators (weight 30%) are based on official data from the International Monetary Fund and World Bank data and refer to debt indicators, credit ratings, access to bank finance/capital markets.

In defining the new ESI index, the Principal Component Analysis (PCA) method was used, which assigns weights to each variable we included in the index formation, giving one index that best represents the values of multiple given variables.

The analysis of the value of the new ESI indicator for the EU countries for 10 years showed that GDP per capita, country risk, carbon intensity, and energy intensity are of high importance for energy security. On the other hand, the share of renewable sources in energy consumption and energy dependence is the least important. The values of the index by country have stable values throughout the observed period, with large differences between countries.

In addition to considering the political, economic, social, and environmental aspects of energy security, the aim of the paper was to consider the correlation between macroeconomic stability and energy security. The results of the research showed that there is a statistically significant difference in the value of the energy security index between the observed groups, that is, the countries with a higher average macroeconomic stability index also record a higher value of the energy security index.

Reliability assessment of the International Index of Energy Security Risk

Even though numerous aspects of energy security have been pointed out, there is still no generally accepted methodology for evaluating energy security. A large number of definitions of energy security and different approaches to measurement open space for detailed analyses and suggestions for improvements, but they are also a suitable field for abuse and manipulation. Although the index construction methodologies are analyzed according to the common index construction steps consisting of data treatment, normalization, weighting, aggregation, and uncertainty analysis, the measurement results among the indices differ. This is because indicators cover different aspects of energy security and include different indicators, and the results are not only different but can often be contradictory.

In this case, the survey results of the reliability of the International Index of Energy Security Risk are presented. The index was developed in 2012 by the Energy Institute of the US Chamber of Commerce and is based on the methodology they previously developed for the Index of US Pat. Energy Security Risk (Index).

The research sample included 25 countries (Australia, Brazil, Canada, China, Denmark, France, Germany, India, Indonesia, Italy, Japan, Mexico, Netherlands, New Zealand, Norway, Poland, Russian Federation, South Africa, South Korea, Spain, Thailand, Turkey, Ukraine, United Kingdom, and United States) during the period from 1980 to 2016. Multiple Regression Analysis, Multicollinearity test plus, Multicollinearity test with Variance Inflation Factors, and Pearson correlation tests were used as methods in the research. The impact of 29 variables on the total variable IESRI was analyzed, as well as the degree of multicollinearity between the variables, both in total, and in the eight groups to which they belong.

The 29 individual variables that make up the given Index are grouped into eight groups: global fuels: global oil reserves, global oil production, global gas reserves, global gas production, global coal reserves, and global coal production; fuel imports: oil import exposure, gas import exposure, coal import exposure, total energy import exposure, and fossil fuel import expenditure per GDP; energy expenditures: energy expenditure intensity, energy expenditure per capita, retail electricity prices, and crude oil prices; price and market volatility: crude oil price volatility, energy expenditure volatility, world oil refinery usage and GDP per capita; energy use intensity: energy consumption per capita, energy intensity, and petroleum intensity; electric power sector: electricity diversity and noncarbon generation; transportation sector: transport energy per capita and transport energy intensity; and environmental: CO_2 emission trend, CO_2 per capita, and CO_2 GDP intensity.

The results of the research showed that the most reliable indicators, which do not exhibit a critical degree of multicollinearity, are variables from the fuel imports group: *Oil Import Exposure, Gas Import Exposure, Coal Import Exposure, Total Energy Import Exposure, and Fossil Fuel Import Expenditure per*

GDP. These variables are indicators of import exposure in general, which indicates the significant importance of dependence on energy imports and the methodological correctness of the setting of each of these variables. If the number of variables that make up the Index is reduced (to increase its consistency and reliability), the mentioned variables should certainly remain part of the Index. On the other hand, for the variables Energy Intensity and Petroleum Intensity, a high correlation was measured, which indicates that they insignificantly contribute to IESRI, and their further participation needs to be reconsidered. Furthermore, although the variables from the transportation sector per capita and transportation energy intensity group have the smallest share in the IESRI variance, both show a high degree of independence and a relatively small degree of overlap, so they should remain part of IESRI.

The analysis of the correlation of the economic group of variables showed a high degree of multicollinearity between GDP-related and indicators from the macroeconomic group and that in further work, we should consider their use in more detail, that is, some should be eliminated. The results show that the variables Energy Expenditures per Capita and Retail Electricity Prices have a negligibly small impact on the formation of IESRI, whereas the impact of the Crude Oil Price Volatility and Energy Expenditure Volatility variables is much higher.

In summary, we observe the biggest problems in the methodological setting of the *International Energy Security Risk Index*. Namely, in their essence, independent variables should be independent, which is not the case here. Contrary to that, all 29 variables show a degree of multicollinearity, which is unacceptably high (>5), since it consequently leads to inaccurate results. It represents inadequate output for decision-making. A high degree of multicollinearity appears when the weight coefficients are set incorrectly, and the obtained p values are, in any case, misleading. Due to the above, the authors propose a new redistribution of weighting coefficients following the results of this study.

Based on the presented results, guidelines for further research can be proposed that should certainly focus on the reengineering of this index. Above all, the emphasis should be on solving the problem of the selection of variables. It should be considered to what extent the selection is necessary, as well as if it is overlapping with the existing variables. Furthermore, when including (or excluding) each variable, it should be assessed whether a change in reliability of the Index occurred. The ultimate goal is to develop a model index, which will contain the variables significant for the observed phenomenon, which are independent (degree of multicollinearity within the allowed limits) while striving to observe the parsimony principle. This implies using only as many variables as necessary (principle of simplicity) but also to use all those variables that are necessary (principle of optimality). Modern trends and efforts to include macroeconomic and environmental indicators in the assessment of energy security, in addition to indicators on energy resources and imports, are sustainable energy transition, but in this case, it has been proven that the choice of variables is wrong, and therefore it is necessary to perform a complete revision. Finally, it is recommendable not to use exclusively mathematical models due to a high level of uncertainty in the sphere of decision-making for energy security and energy transition in general, but to examine the possibility of using fuzzy logic and other inexact systems of reasoning.

Energy security and policy making

From a practical point of view, energy security is becoming an increasingly important issue for each country, but understanding and practical monitoring of energy security are often insufficiently methodologically based, and the estimates thus obtained can be potentially dangerous. Also, from previous research on the essence of energy security, it becomes quite clear that it cannot be viewed separately from the specific characteristics of each country and the moment in which it is determined. Thus, determined national approach to defining, conceiving, and planning an energy secure future creates certain differences between countries while losing a unified approach to energy policy, which is especially evident in the example of EU countries. Every country in the EU (and beyond) is striving to find a way to ensure an energy-secure present and future while leaving less and less room to respect the previously accepted general directions of development. Each country individually strives to be as energy-dependent as possible, because this clearly leads to the development of all other forms of dependence, which are usually unfavorable in the long run, and often lead to the development of geopolitically conflict situations.

Consideration of energy security is completely inseparable from the consideration of national security as a whole. Namely, achieving the country's energy security without undermining its national security as a whole is certainly in the focus of interest.

There are several reasons why the issue of energy security is a concept that depends on the specifics of each country. First of all, there is an uneven distribution of energy resources in the world and within one country. Also, there is unequal access to energy resources, unequal demand in different periods and seasons. Furthermore, countries have different visions of their own economic, environmental, and social future, different geopolitical priorities, and plans for the future. In such a situation, any methodology for measuring energy security can only give an approximate insight into the situation in a given country at a given moment. Finally, the large and different impact of changes in energy prices on energy security is very clear. Namely, high energy prices certainly have a direct impact on economic development, but even in these cases, a country's economy does not reduce energy consumption proportionally. On the other hand, low energy prices affect the increase of consumption and especially the reduction of investments in renewable energy sources and energy-efficient technologies.

Therefore, regardless of the definition used, energy security today is defined at the national level, following one's vision of the present and plans for the future. The efforts of each country to achieve and maintain energy self-sufficiency, as well as the development and preservation of an energy system that is resistant to disturbances—are becoming an imperative of energy policy in all countries that strive to preserve national security as a whole. Therefore, the stability of the energy system, its vulnerability, resilience, and self-sufficiency should be the basic parameters of energy security. In addition, it is necessary to take into account the fact that energy security is a dynamic category, because at different times there can be different priorities in the same country. Based on all that has been previously said, we can conclude that stability in the field of energy security is much more significant than the measured size itself, as well as the economic and environmental costs that arise in this regard.

In addition, the relationship that exists between the parameters that affect energy security has been clearly defined in a large number of studies as more important than the quantities in which the values of these parameters are expressed. Therefore, knowing the relationship between economic growth, sustainable development, and energy security is of great importance for the future. Most of the previous research cites as one of the conclusions the need for continuous review of existing definitions and proposals for new solutions. Namely, energy security is not a new concept, but it certainly requires a new way of observing, defining, and measuring.

A high degree of uncertainty is a problem faced by all decision-makers in the modern world, especially when it comes to energy policy. When a large amount of data is added to the equation, it is absolutely clear that a large number of challenges open up for decision-makers. There is a lot of data, but their accuracy and credibility are often questionable. Modern information technologies enable the application of numerous methods and techniques; however, all the advantages simultaneously carry significant deception and can lead to making some intentional or accidental decisions with far-reaching consequences. All of the above points to caution, multidisciplinarity, and objectivity, with a realistic view of the data and modern processing methods, in the process of energy security modeling.

In such a complex equation with too many unknowns, it is difficult to make decisions. However, due to the high importance of energy security in the modern world, decisions must be made in the given circumstances, with the data available to decision-makers. Energy decisions often have to be made relatively quickly and in line with geopolitical changes, changes in financial markets, and with the clear knowledge that these decisions are far-reaching and can sometimes lead to serious unintended consequences, most notably internal unrest, diplomatic, and armed conflicts. The main challenges that will have an impact on energy security in the future are:

- Complex geopolitical changes;
- Unequal distribution of energy resources and proven reserves;
- Increase in energy consumption;
- Undefined issue of justification of investing in renewable energy;
- Disagreements over the use of nuclear energy;
- The need to reintroduce coal mining;
- Insufficient and inadequate infrastructure;
- Realistic assessment of the formation and impact of energy prices;
- Educated and trained people;
- Management of natural resources in underdeveloped countries;
- The problem of availability of rare metals;
- Wars, riots, terrorism and
- Cyber-attacks.

Energy security is part of the national security system. The issue of energy security and national stability is complex and exposed to a large number of influences, many of which are geopolitical, with a high degree of unpredictability and variability. The problem of decision-making is further complicated by the unclear nature of the variables that must be taken into account in strategic planning, the undefined direction and intensity of their impact, their complexity, and the impossibility of exact quantification.

Case study

Sustainable energy development: How can the tension between energy security and energy transition be measured and managed in South Africa? Journal of Cleaner Production, Volume 205, 2018, Pages 738–753. https://www.sciencedirect.com/science/article/abs/pii/S0959652618325393.

References

Akhtaruzzaman, M., Boubaker, S., Chiah, M., & Zhong, A. (2021). COVID – 19 and oil price risk exposure. *Finance Research Letters*, *42*, 101882.

Akins, J. E. (1973). The oil crisis: This time the wolf is here. *Foreign Affairs*, *51*(3), 462–490.

Ang, B. W., Chong, W. L., & Ng, T. S. (2015). Energy security: Definitions, dimensions and indexes. *Renewable and Sustainable Energy Reviews*, *42*, 1077–1093.

Augustis, J., Krikštolaitis, R., Martišauskas, L., Pečiulytė, S., & Žutautaitė, I. (2017). integrated energy security assessment. *Energy*, *138*, 890–901.

Aven, T. (2016). Risk assessment and risk management: Review of recent advances on their foundation. *European Journal of Operational Research*, *253*(1), 1–13.

Azzuni, A., & Breyer, C. (2018). Energy security and energy storage technologies. *Energy Procedia*, *155*, 237–258.

Boyko, V., Rudnichenko, N., Kramskoy, S., Hrechukha, Y., & Shibaeva, N. (2017). Concept implementation of decision support software for the risk management of complex technical system. In N. Shakhovska (Ed.), *Vol. 512. Advances in intelligent systems and computing. Advances in intelligent systems and computing*. Cham: Springer.

Cherp, A., & Jewell, J. (2014). The concept of energy security: Beyond the four As. *Energy Policy*, *75*, 415–421.

Filipović, S., Radovanović, M., & Golušin, V. (2018). Macroeconomic and political aspects of energy security. *Renewable and Sustainable Energy Reviews*, *98*, 428–435.

Franc-Dąbrowska, J., Mądra-Sawicka, M., & Milewska, A. (2021). Energy sector risk and cost of capital assessment—Companies and investors perspective. *Energies*, *14*, 1613.

Gasseer, P. (2020). A review on energy security indices to compare country performances. *Energy Policy*, *139*, 111339.

Hammoudeh, S., Makni, K., Ben-Salha, O., & Ajmi, A. N. (2021). Distributional predictability between oil prices and renewable energy stocks: Is there a role for the COVID-19 pandemic? *Energy Economics*, *103*, 105512.

Iannou, A., Angus, A., & Brennan, F. (2017). Risk-based methods for sustainable energy system planning: A review. *Renewable and Sustainable Energy Reviews*, *74*, 602–615.

Index of U.S. Energy Security Risk. (2018 Edition). *US Chamber of Commerce*. Global Energy Institute. https://www.globalenergyinstitute.org/index-us-energy-security-risk-2018-edition.

Kester, J. (2021). The scare behind energy security: Four conceptualisations of scarcity and a never-ending search for abundance. *Journal of International Relations and Development*. Published May 2021.

Malkova, T. B., & Malkov, A. V. (2018). Practical method of risk assessment in the energy sector under uncertainty. *Russian Journal of Industrial Economics*, *11*(1), 63–69.

Matsumoto, K., & Shiraki, H. (2018). Energy security performance in Japan under different socioeconomic and energy conditions. *Renewable and Sustainable Energy Reviews*, *90*, 391–401.

Murshed, M., & Tanha, M. M. (2021). Oil price shocks and renewable energy transition: Empirical evidence from net oil-importing South Asian economies. *Energy, Ecology and Environment, 6*, 183–203.
Narula, I., Reddy, B. S., Pachauri, S., & Dev, S. M. (2017). Sustainable energy security for India: An assessment of the energy supply sub-system. *Energy Policy, 103*, 127–144.
Papież, M., Śmiech, S., & Frodyma, K. (2019). Effects of renewable energy sector development on electricity consumption—Growth nexus in the European Union. *Renewable and Sustainable Energy Reviews, 113*, 109276.
Paravantis, J., Kontoulis, N., Ballis, A., Tsirigotis, D., & Dourmas, V. (2018). A geopolitical review of definitions, dimensions and indicators of energy security. In *2018 9th international conference on information, intelligence, systems and applications (IISA). IEEE* (pp. 1–8).
Podbregar, I., Šimić, G., Radovanović, M., Filipović, S., & Šprajc, P. (2020). International energy security risk index—Analysis of the methodological settings. *Energies, 13*(12), 3234.
Sovacool, B. K., & Mukherjee, I. (2011). Conceptualizing and measuring energy security: A synthesized approach. *Energy, 36*(8), 5343–5355.
Radovanović, M., Filipović, S., & Golušin, V. (2018). Geo-economic approach to energy security measurement—Principal component analysis. *Renewable and Sustainable Energy Reviews, 82*(Part 2), 1691–1700.
Report: A quest for energy security in the 21st century. Asia Pacific Energy Researchz Centre, Tokyo, Japan. (2007).
Report: Cyber security in the energy sector. Recommendations for the European Commission on a European strategic framework and potential Future legislative acts for the energy sector. (2017).
Report: World energy assessment: Energy and the challenge of sustainability, UNDP. (2015).
Spada, M., & Burgherr, P. (2020). Comparative risk assessment for fossil energy chains using bayesian model averaging. *Energies, 13*(2), 295.
Sýkora, M., Marková, J., & Diamantidis, D. (2018). Bayesian network application for the risk assessment of existing energy production units. *Reliability Engineering & System Safety, 169*, 312–320.
Thorbecke, W. (2019). How oil prices affect East and Southeast Asian economies: Evidence from financial markets and implications for energy security. *Energy Policy, 128*, 628–638.
Tian, W., Heo, Y., Wilde, P., Li, Z., Yan, D., Park, C. S., et al. (2018). A review of uncertainty analysis in building energy assessment. *Renewable and Sustainable Energy Reviews, 93*, 285–301.
Winzer, C. (2012). Conceptualizing energy security. *Energy Policy, 46*, 36–48.
Wang, Q., & Zhou, K. (2017). A framework for evaluating global national energy security. *Applied Energy, 188*, 19–31.
Wieczorek-Kosmala, M. (2020). Weather risk management in energy sector: The polish case. *Energies, 13*, 945.
Wang, K., Mao, Y., Chen, J., & Yu, S. (2018). The optimal research and development portfolio of low-carbon energy technologies: A study of China. *Journal of Cleaner Production, 176*, 1065–1077.
Yergin, D. (1988). Energy security in the 1990s. *Foregin Affairs, 67*, 110–132.

Index

Note: Page numbers followed by *f* indicate figures.

D

E

F

T

Kolibri 360 GmbH
Ferdinand-Jühlke-Straße 7
99095 Erfurt, Deutschland
produktsicherheit@kolibri360.de